Alfred James Swinburne

Picture Logic

An Attempt to Popularise the Science of Reasoning

Alfred James Swinburne

Picture Logic
An Attempt to Popularise the Science of Reasoning

ISBN/EAN: 9783744723961

Printed in Europe, USA, Canada, Australia, Japan

Cover: Foto ©berggeist007 / pixelio.de

More available books at **www.hansebooks.com**

PICTURE LOGIC

AN ATTEMPT TO POPULARISE THE SCIENCE OF
REASONING BY THE COMBINATION OF HUMOROUS PICTURES WITH
EXAMPLES OF REASONING TAKEN FROM DAILY LIFE

BY

ALFRED JAMES SWINBURNE, B.A.

QUEEN'S COLLEGE, OXFORD

The Lion of Human Understanding in the tangle of Logical
Knots assisted by the Mouse of Illustration

WITH ORIGINAL ILLUSTRATIONS FROM DRAWINGS BY THE
AUTHOR ENGRAVED ON WOOD BY G. PEARSON

FIFTH EDITION

LONDON

LONGMANS, GREEN, AND CO.

AND NEW YORK : 15 EAST 16th STREET

1887

INTRODUCTION.

It was at the beginning of a certain Long Vacation when my father sent for me and delivered himself of the following remarks : ' My son, your scores at cricket, your racquets, your prowess in the hunting-field and in your college steeple-chases, your numberless invitations and popularity, to you doubtless appear all that can be desired ; to me, Sir, they are nothing—nay more—they are even positively harmful, seeing that by their fascinating brightness men are blinded to all sense of their true interests and aim— viz., to secure their degree as soon as possible with a view to a start in life.' Upon my replying to my father to the effect that every allowance was to be made for him—as having left college five-and-twenty years—if, as in the present instance, he manifested lamentable ignorance of the whole state of the University at the present day, and that his milk-and-water reading man would certainly be regarded with loathing and abhorrence by all ' our fellows ' and all the best men at Oxford, and consequently, sinking into obscurity, would be ruined for life, and upon my making many other similar assertions, my father, with much warmth, commanded me to be silent, and then asked me if I expected I was to live a life of slothful ease, because I was a rich man's son ; with several other questions which were not meant to be answered ; finally becoming so excited

as to refer me to his own university career, a subject which
he quickly dropped, remembering how often he had told me
stories of his undergraduate days before I was sent to college. The result was that I was ordered to select a tutor
for two months in the Long Vacation and pass my modera-
tions in the following term, or for ever be condemned to
the backless slippery heights of office stools. The awful
thought of ' wasting my sweetness ' and withering in such
a dry and uncongenial soil nerved me for a desperate effort.
Of a restless and excitable disposition I was for some time
after haunted by dreams of men with pens in their ears,
and ledgers with columns of figures to add, so lofty that
their bases were on the earth while their summits were lost
in the clouds. I never could do mathematics—not that I
was quick at any work—even my mother allowed this, for
she wrote to my tutor for matriculation to the effect that
' our dear Douglas had manifested symptoms of future
greatness, when a child, and still possessed *remarkable*
ability, if it could only be drawn out; but alas! there was
a want of application, especially in his mathematics.' I
therefore determined to take up Logic as a substitute for
Mathematics, and wrote to inform my tutor that I should
only want help in this subject. He selected a charming
spot on the north coast of Devon and we met there. He
had one other pupil—a very quiet youth and, as it seemed
to me, very clever, my fear of whom was heightened con-
siderably when I learnt that he had intended to try for a
class, but, finding his books in a very imperfect state, was
content with passing, though determined not to miss that.
The awe with which this piece of information filled me I
never succeeded in quite shaking off, though I liked him
very much afterwards. He always seemed to me a sort of
half-way house between Mr. Practical and myself—the idea
of any one knowing more than Mr. Practical was an idea

that never for a moment entered my head Old Prac'
(as we called him afterwards) had such a smooth, comfort-
able way of settling any difficulties I proposed—so reassur-
ing that I verily believe if he had told me that the best way
to learn the art of diving and remaining under for a long
time was to tie a heavy stone round your neck and get some
one to push you in, I should have tried it. His last words
the first night were—' Logic to-morrow.'

It is needless to say my sleep was much disturbed that
night with anticipations and forebodings. What was this
new and strange study ? Had I not always heard men
speak of its difficulty ? How if the momentous question,
' Was I possessed of a " turn " for Logic ? ' should be answered in the negative ; and I fell asleep to dream of
mysterious figures, numbers, and symbols on the one hand
pitted against the mocking forms of clerks, managers, and
office boys on the other.

CONTENTS.

———

a

LIST OF ILLUSTRATIONS.

—✦—

PICTURE LOGIC.

CHAPTER 1.

WHAT IS SCIENCE?

NEXT morning Mr. Practical assumed a grave look and began : 'There is no lack of treatises on Logic, but there is a lack of people who understand them. It is the custom of Passmen to attempt to learn by heart a great deal of matter they do not in the least comprehend, without attempting to realise the meaning, and so fixing it in the memory. A little understood is better than a volume learnt by heart. I shall not expect you to remember anything you do not understand ; nor shall I ever make use of instances other than those of everyday life ; and if my illustrations be too familiar to appear scientific, my excuse will be that I wish to bring home to you what I say, that you may realise and appreciate and so remember its meaning, as a man who has swum two miles, or been ill eight hours on a boat, has no difficulty in remembering the force of the expressions " long swim " or

" unpleasant cruise."   The neat and concise phrases
you meet with in your treatises on Logic are decep-
tive—like the ease of a good skater or runner, they
are the results of hard work, and you might as well
expect to get that ease at once without practice, as
imagine that by learning off those phrases you have
mastered Logic.   Therefore, pass over all words and

Nonplussed.

instances that you do not understand as you read
your treatises side by side with my attempts at explanation.'

It would be difficult to express my sensation of
relief at these words ; for already I had opened my
new 'Elements of Logic,' and, flushed with hope,
peeped at Chapter I. ; but, confronted by the very
first word, 'Psychology,' I felt as staggered and faint

as when, burning to learn the noble art of self-defence, I put on the gloves with the noted Punisher, and he knocked me down the very first blow of the first lesson. 'Courage,' I muttered, and pushing on, saw something about 'analysing phenomena' and their 'mutual relations' and 'the mode of their generation;' and after this, I became unconscious, until I was aroused by a shout of breakfast, and found myself sitting staring blankly at my book with ideas of chemists, newspaper accounts of 'a strange phenomenon,' aunts and cousins hopelessly complicated, and letters in the 'Field' about the breeding of fowls, crowding in wild confusion across my mind.

'Tell me,' resumed my tutor, 'what you mean by Logic, Dyver.' 'Logic,' replied he, 'is the science and art of reasoning, or the science of the conditions upon which correct thinking depends, and the art of attaining to correct thinking, or the science of the formal laws of thought—and other definitions might be given.'

Observing my look of despair, Mr. Practical went on : 'Logic seems to be a science and an art, but if I asked you what is a serpent, and you replied "a reptile," this would be no answer if I did not know what "reptile" meant. "This is no answer, thou unfeeling man!" the beginner might well exclaim, were I to rest content with defining Logic as the science and art of reasoning. If Logic is a science

and art let us leave Logic alone for the present, and
when we have grappled with these two mysterious
shadows Science and Art, return to our definitions to
find that the apparently different accounts of Logic
are only various ways of expressing the same thing.
Science is knowledge—but this is no explanation
until we know what knowledge is. First, then, let
us try and explain what knowledge is.

Imagine the whole world (*i.e.* everything that
exists—including of course the stars and the heavens
and yourself) to be divided into two parts ; on the one
hand yourself, a being with faculties of observation ;
and, on the other hand, everything but yourself, objects which fall under that observation. And you must
remember that the observing powers can be turned
back upon yourself, so that in this sense you yourself
may be said at the same time to belong to both parts,
as being the person observing and the object observed.
You don't follow, Destrawney ? It will be clearer
presently. Now all that vast number of things which
form the part which is not yourself have, so to speak,
peculiar features and differences, qualities, or attributes, or whatever you choose to call them. What I
mean is that they affect you differently as you hear,
see, smell, touch, or taste them. Take one or two
instances, a star—a tree—a pond. Each of these
objects has its peculiarities or attributes. The star
shines, the tree blooms, the pond is stagnant. You
are differently affected as you glance from the one

to the other. And this employment of the senses given you by nature is knowledge. The world around you is teeming with attributes, qualities, or peculiarities; particular facts you may call them or phenomena (from the Greek φαίνομαι, I appear or show myself), and the noting or acceptation of these particular facts by the senses is knowledge. The lion roars, the rain falls, the flame rises, &c., and when I say so I prove the existence of knowledge in me.'

'I think I have some vague idea of your meaning,' said I, 'but I always thought knowledge was something grander—more difficult somehow.'

'Of course,' he went on, 'I only take the simplest instances of the employment of our senses on the world around us. It must be borne in mind that each thing included under that term has an infinite number of such attributes or phenomena. Take the moon. That it is round and mountainous, &c., we know, but there are a countless number of peculiarities of the moon that we do not know anything about, and the advance of knowledge only means the observation of, or direction of our senses to, attributes or phenomena not yet discovered or known to exist. The man who first applied steam to locomotion had noticed an attribute in steam hitherto unobserved, viz. its elasticity. Knowledge, then, is the application of our senses to the phenomena or attributes of things around us. But it is something more than the mere five senses that man employs.

The brute creation can hear, see, smell, touch, and
taste; but man has a higher power as well—that we
may call reason; a faculty by which he can gather
and group together particular facts and form universal ideas and propositions.   Not only are we
aware of the presence of "this horse," "that draper,"
"yonder mountain;" but we can close our eyes and
picture to ourselves the image of a horse generally
that is not any horse in particular; and so with the
draper and the mountain.   And not only can we
say "this horse is four-legged," "that draper is
weak," "yonder mountain attracts the clouds;"
but we can also say, "all horses are four-legged,"
"all drapers are weak," "all mountains attract the
clouds," supposing our observations to have been
sufficient to ground these universal propositions
upon.   And notice the signs of the two different
propositions.   The sign of a particular proposition is
generally "this" or "that" or "some";[1] of a universal, "all" or "no."'

I could not help thinking how learned I was
becoming, and how easily I could now refute my
father if he ever again dared to argue with me.

'Now, this second kind of knowledge we call
universal knowledge as opposed to particular knowledge, and the results attained to are called generali-

[1] 'This' or 'that' may in another sense be regarded as the signs of
universal propositions, when they mean the 'whole of this' or the
'whole of that.'—See p. 74.

sations, uniformities, universal propositions, or laws, or principles, as opposed to the particular facts from which they are derived.'

'Need we remember all those names?' I gasped.

'Never mind them for the present,' he continued. 'Remember that these universal propositions derived from particular facts are called science, and thus science is universal knowledge (where knowledge is used for the results attained as well as the process of attaining them). And science is divided into branches according to the nature of the objects with which it is concerned, and each branch is spoken of as a separate science. Thus, if I gather universal truths from particular facts observed about the stars, the science is astronomy; and if about the earth, geology; and if about trees, botany. For instance, I observe *this* tree buds in the spring, and *that* and *that*, and so on until at last I lay aside my particulars, and assert once for all that "all trees bud in the spring," and this proposition forms a part of the science of botany. Now, gather your universal truths from particular facts observed about thought and the science is Logic.'

At this triumphant flourish Dyver, who had for some time exhibited signs of restlessness, could restrain himself no longer, but burst in with his difficulties. 'Firstly,' cried he, 'what do you mean by thought? and secondly, why should men care to draw these generalisations that you speak of?'

'Of thought more anon,' replied our tutor; 'as

to the generalisations, you must recall to your memory
the division above mentioned, yourself or mind, on
the one hand, and all external phenomena or matter
on the other; and on the side of mind or yourself.
You will find implanted by nature an inclination to
gather together the like, to classify, and arrange,
and group similar phenomena together—a yearning
after generalisation—more conspicuous in women
than in men, because in women the natural impulses
are less restrained by reason. But of this hereafter;
for the present recollect what we have said, that
knowledge was an employment of the senses upon
external phenomena; that knowledge was of two
kinds, universal and particular; and that universal
knowledge is called science, and that the branches
of science are named from the objects with which
they have to do; and the science of thought is Logic.
Of art we must speak to-morrow.'

That day we went for a lovely walk, Dyver seem-
ing buried in meditations of future knots for logical
solution, and whether it was that I over-tired myself,
or whether it was through working a brain usually
idle, I don't know, but I had a remarkable dream
that night. I saw an old gentleman who had been
my house-master at Eton, standing at the foot of a
range of hills, and busily engaged in picking up
things, and putting something now and then into
a huge preserving jar that was at his side. On
approaching nearer I heard him muttering some-

SCIENCE PRESERVED IN JARS.

thing over and over again—about knowledge and triumphs: 'Rather tiring work, isn't it?' I ventured to ask. 'What triumphs do you refer to?' 'Why, the triumphs of science, young man,' he sharply retorted. 'Don't you call it a triumph to arrange, group, and classify yonder untidy wilderness of particulars? to detect similarities and form universal propositions, or uniformities, or laws, and lay by these laws in jars for future use? to introduce neatness and order in that tangled medley before you, that bewildering chaos of confusion?' 'Most certainly I do, but I don't quite see how you mean. I don't pretend to be clever, though.'

'Keep silence, and observe the process. I'll take a very simple uniformity to make it plain to you.' He then picked up a stone and let go of it, and it fell to the ground. '*This* body tends to fall to the earth,' said he; and repeated the operation with a book, a chair, a bottle, a purse, and several other things, each time repeating the formula, 'This body tends to fall to the earth,' much in the same way as a man repeats the responses in church. Then he wrote something on a slip of paper, and handed it to me. It was—'All bodies tend to fall to the earth (uniformity or law of gravity).' 'There,' said he, 'that's the process; of course simple uniformities like these were observed long ago. I'm engaged in much more abstruse and complicated work now, but the process is precisely

the same.' He slipped the paper into one of the
jars already filled, corked the jar carefully down, and
patting it affectionately, continued, 'You see the
advantage. What a saving of time and trouble, for
we need not look every time at our stone, book,
purse, &c., *i.e.* the particulars, but we have our com-
prehensive law once for all, and we can dispense
with all further experiments with the particular
facts. All knowledge, indeed, is power, and this is
as true in complicated instances but less obvious
than here. You hold in your hand a brittle orna-
ment; without any physical effort your knowledge
gives you the power to break it, for you know that
if you let it go it will fall to the earth and break,
because you have drawn the law of gravity from the
observation of particular facts.'

'Quite so,' said I, growing a little weary; 'but
excuse me, what is that bump on your forehead?'

'That,' said he, 'is the bump of generalisation,
and it indicates an innate desire in man to observe
similarities, and group and gather particulars, and in
virtue of the possession of similar qualities, and so
make laws. I've great trouble in preventing people,
especially women, from giving way to this impulse
too much. Men have what they call "hobbies," and
women "jump at conclusions."' As I was musing
upon this, and thinking how I would enlighten their
dark minds at home, I was recalled to myself by
an angry shout from Mr. Science (for that he told

me was his name). Turning my eyes in the direction
of his shout, I saw an old woman flying through the
air with a large comet under her right arm. She
clung tightly to it as if for support. 'Ho, there,
where are you going to with that comet?' cried he.
'O please, Mr. Science,' said she, 'I've got a law at
home that "all comets are signs of war," and as mother

Superstition.

made the law without seein' no pertiklers, we thought
we'd collect pertiklers arterwards.' 'Be off,' yelled
the old man in a perfect frenzy of rage, 'how dare
you make laws without a sufficient number of particulars to warrant your conduct. Be off, I say, you
old baggage, and drop that comet this moment!' I
covered my face lest he should see my amusement at
his ferocious gesticulations, when the old woman fled,
still clasping the comet in her arms.

'You see those two hills beyond,' he said, turning to me again; 'we find it hard work there. We've made most progress in mathematics and astronomy. We can get our particulars separately there; but those hills are composed of the particulars of chemistry and human conduct, and they're very difficult to get at. And as one of the chief advantages oi science is its prediction (for when we know that all steam is elastic, we can predict that any particular steam will be elastic), we lose much by not having its aid in the arrangement of these particulars.'

'One might almost compare your work to mining,' said I, 'you seek for the gold of truth in things, and when you have found it you can throw away the earth or particular facts.' This seemed to me a splendid simile, and I only wish Dyver could have heard it, but the old man took no notice of it at all.

'It's very hard to find uniformities in things,' he continued, 'when you can't get them separately. Supposing gunpowder was always found with some other substance, and you couldn't separate them, how would you know whether it was the gunpowder or the other substance which possessed the quality of exploding when a spark was applied? So you can't get your uniformities or laws, and science is backward there.'

'Or supposing,' said I, 'two men always visited your house together, and at the sight of them the dog always exploded into barking, unless you could separate or isolate the men, you couldn't make any law as to which of the men the dog was barking at.'

'Yes,' said he, smiling contemptuously, 'that would do, but a better instance is' . . . . and he went on talking about crystals and double refractory powers till the whole scene changed, and I found myself climbing up, as of yore, to get a nearer view of the beautiful glass lustres of the chandelier at home, and upsetting the ink, and being called by my mother a refractory child, for that's all I knew about 'crystals' and 'refractory powers.'

CHAPTER II.

WHAT IS ART?

NEXT morning I arose, wondering at my dream, in which I had seen much more than I had heard from Mr. Practical, and accounting for the prodigy by supposing that all the scraps I had read at odd times in our domestic British Encyclopædia had been aroused from their sleep in my memory by my late efforts to think.

'If science,' resumed Mr. Practical, 'may be called universal knowledge derived from particular facts, art is the application of that universal knowledge to particular facts, and we may speak of the arts as we do of the sciences, meaning thereby branches of one great art and branches of one great science. By science I establish the law, for instance, that "all trees bud in the spring." Now supposing I were able to produce a tree, I should say to myself, " Let me see, a tree is to be produced ; first of all it must bud in the spring, for unless it did so it would not be a perfect tree, and then it must be something else, and so on; and as a sculptor complies with orders in a contract, so must I obey laws in the production of this tree, and this is what we call the application of universal knowledge to particular facts. And this is why the laws of a science are also called conditions, because without complying with them we cannot produce a correct instance of a particular : e.g., from observation we enunciate the law, "all water has drowning properties." Now this is a condition upon which water's being water depends, for if a man pointed to some water and said, " This water has no drowning properties," we should say, " My good sir, this cannot be water, then, for it does not comply with the conditions required to constitute water." Now art is the application of universal laws in this way to particular facts, strictly speaking, in production, but also in criticism.'

' I don't quite understand,' said I.

' By production, I mean the process of making something, as a picture, statue, or house ; by criticism the process of testing the genuineness, or right to its name, of something, as a star, tree, or man. In either case we apply universal knowledge to particular facts, and the process is called art.'

Dyver, looking very subtle, here enquired whether science was the *process* of attaining to universal knowledge, or the *knowledge* so attained, for art seemed to be the process only, and Mr. Practical replied that it was difficult to say ; he believed science was used for both, perhaps on the whole it was more often used for the knowledge attained, but it made very little difference to the broad distinction between science and art, and that was all he wanted to show.

'To give you an illustration of what I mean by science and art, I shall make a rough sketch with my pencil,' he continued, scarcely a bit discomposed by the penetrating subtlety of his senior pupil.

The Beginning of Science.

'Here we have an aged admirer of scenery. Night after night he gazes on the sunset from his rocky seat. He observes certain points possessed in common by each sunset (to speak of sunsets as distinct particulars). The clouds may vary, but there are certain particular rays of light he sees every time he sees the sun set. He thinks to himself, " I observe these sunsets agree in certain features—they all have certain rays of light, &c. These common elements, or uniformities, I will gather together and

make laws of them for the guidance of my boy, who
has shown a taste for painting." '

The Beginning of Art.

So he draws up these common elements in the
form of laws, and teaches them to his boy. 'All

correct sunsets have such and such rays,' &c. And the boy grows up, and carries out in a particular instance the laws his father drew from particular instances, by painting a picture of a particular sunset which obeys the laws his father drew; and if it obeys all of them, it is said to be a very good painting and a perfect work of art, and it hangs in some national gallery. The first two illustrations here would come under science; the last, art.' While I was laughing at the family likeness between the father and son, 'Could you give us an instance of criticism as well?' asked Dyver; 'the above is obviously production.'

'I suppose it might fairly be called art to apply universal laws to particulars in the way of testing their truth or genuineness, and this would be criticism. A judge of pictures looks at the sunset above depicted, and tests its truth by calling up one after another the laws gathered from the observation of particular sunsets, and seeing whether the picture before him complies with these laws. And this is what is meant by saying that a picture presents to us the essential, as well as any accidental, features of its original.'

'But is it the same,' said I, 'with the science and art of riding, or swimming, or skating—those sixpenny sciences, I mean, that you buy in yellow covers with pictures on?' 'Precisely,' replied he, smiling at my earnestness; 'the particulars of swimming are the particular movements of the limbs by this or that swimmer. These are observed by the man of science—their uniformities or common elements constituting laws or conditions—and he enunciates them : "All correct swimmers spread their limbs in such a way," and several other laws. Should he feel disposed to learn swimming himself, and enter the water, attempting to spread his limbs according to these laws, he would be employing art. And the case is similar with shooting, or riding, or skating. Wherever you draw universal knowledge from particular facts you have science (the name varying with the subject matter), and wherever you apply your universal laws to particulars, again you have art (the name varying in the same way), and science and art applied to thought share the name of Logic. Thus the beginning of science and art taken together is the same as their end, *viz.* particular facts, as you see in the case of the painter, the particular sunsets you see in fig. i. and fig. iii. are the beginning and the end.'

All of a sudden Dyver clutched his pencil convulsively, and became so engrossed in his note-book that I very nearly burst out laughing ; it reminded me so of the way we used to clasp our lexicons at school, when the master returned unexpectedly, after leaving us alone for a few moments. 'Might not this figure,' he presently asked, 'represent science and art? it has just occurred to me it might.' This was

the figure, and Mr. Practical was highly pleased
with it; and the more so because he told us universal knowledge is always regarded as something

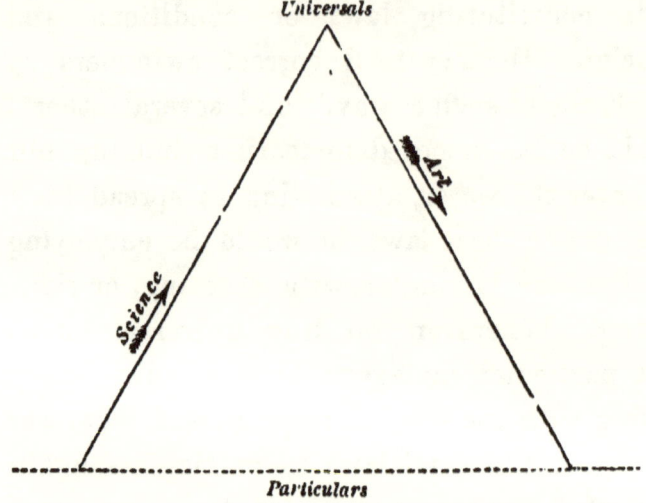

higher than particular—we 'ascend' to it and *deduce*
from it—and we talk of particulars *de*pending upon
or hanging from laws or conditions.

CHAPTER III.

LOGIC IS A SCIENCE AND AN ART.

'YESTERDAY,' continued Mr. Practical, 'we saw what science and art were. If Logic claims to be considered a branch of science and art, we must not be content until we have heard proof that Logic gathers laws from particular facts, and applies laws to particular facts.

'We have already observed that the eye of the soul, so to speak, can be turned back upon itself—and we can contemplate our very thoughts. And in this field or sphere of thought we find particulars to work upon, and Logic grounds upon these particular thoughts uniformities, conditions, or laws in its capacity of science, and applies these uniformities, conditions, or laws to practice, in its capacity of art.'

'But thought,' said I, 'is so vague and difficult, please make the whole thing clearer. I think I begin to understand about some of it.'

'In a subsequent lecture I shall endeavour to discuss thought more thoroughly—showing how all thought is comparison. To take, for the present, a common-sense view of the matter, thought may be either unexpressed or expressed. You may be sitting next to a lady at dinner and you may think " This lady is very ugly," this, we trust, would be an unexpressed thought. Or you may think " The weather is lovely" and turning to the lady you may express this thought, and it becomes a remark. Again, you may in your mind form a more complicated kind of thought. You may think, " All cucumber is indigestible ; this is cucumber; therefore it is indigestible," a thought unexpressed also. Or you may think, " All young ladies are charming companions ; my neighbour is a young lady; therefore my neighbour is a charming companion," a thought you may express to her, and it becomes an inference. Now it is obvious that there is an infinite number of remarks and inferences that one might make, and we shall regard these remarks and inferences as thought for the present—for they are thought viewed from the outside of the lips, so to speak, or Homer's ἕρκος ὀδόντων. Now, given a number of these various remarks and inferences, you will find some correct, some incorrect. Here, then, is work for science—to collect a number of correct remarks or inferences (as they are called), find out their uniformities, their common elements (for something in common they must have, or why were they called by the same name ?), erect these into laws and call these laws the principles or conditions upon which correct remarks or inferences depend—precisely as the botanist did with his trees, the astronomer with his stars, and the old man with the sunsets. Here, too, is work for art, for the principles or conditions can be employed either in the production of particular correct remarks and inferences or in the criticism of the particular remarks and inferences of others. And Logic is the name given to the science and art, for these remarks and inferences represent thoughts.'

'How do you know when a remark or inference is correct?' asked Dyver.

'This is just what Logic would tell us, by giving us the laws required to be complied with by any remark or inference that is entitled to the name of "correct." As a matter of fact, we generally know by instinct and without any extraneous aid whether any remark or inference is correct or not. Particular correct thoughts were distinguished from particular incorrect thoughts before Logic was thought of, but Logic gives us the why and wherefore, and enables us consciously to detect errors they unconsciously detected before.'

'I am afraid I don't understand these last words,' I remarked.

'Never mind. All you need recollect is this —certain thoughts are called correct among men and certain thoughts incorrect. You step into a third class carriage. It is full of costermongers and illiterate persons. The man next to the window makes a remark, "Deep rivers is allers shallow." A look of wonder steals slowly over every face. The man proceeds to an inference. "Balloons is the most riskiest travellin', and I'm a goin' by the Underground Railway; so I think its pretty sartain I'm a goin' by the most riskiest travellin'." There is a hoarse laugh from all, and they tell the guard to look after that man as a lunatic, one with reason or thoughts perverted, for it wants no logic or cleverness to detect an incorrect thought. You return to your first class carriage, and a friend tells you there that what those men unconsciously did Logic enables us consciously to do, for Logic gathers the common elements of correct thoughts and making laws of them gives us principles for guidance in the production of thoughts ourselves and the criticism of thoughts in the case of others.'

'Could you tell us any of these laws?' asked Dyver.

'There are three primary laws to which all the laws of thought may be reduced; and you only need know their names for the present. They are—

I. The Law of Identity (whatever is, is).
II. The Law of Contradiction (a thing cannot both be and not be).
III. The Law of Excluded Middle (a thing must either be or not be).

Simple and even absurd they appear at first sight, but still they are laws obeyed, though unconsciously, by millions of men every day; and until you realise their meaning, respect them for their universality: I shall devote more time to them afterwards.'

'Do you mean to say,' indignantly demanded Dyver, 'that by learning these mere truisms—for that's what they seem to me to be—that we are helped at all in the attainment of correct thoughts, and the avoidance of error? Suppose I want to invest some money in the Peruvian Stocks. The only question that frightens me is, "Is the guano adulterated?" I consult a logician as the wisest of men. His oracular response is, "In reference to this matter, I can tell you one thing. If it *is* adulterated, it *is* adulterated; and furthermore, since you seem much excited about it, I will proceed to the disclosure, 'that it *can't* be both adulterated and not adulterated' (at least, in the same place and time); while, on the other hand, error might be incurred, were you not to keep well in mind the fact that guano *must* either be adulterated or not adulterated.' What should I gain by such information? Should I not, in a fit of anger at his cold, dull certainties, probably invest, and lose my money? and would not this be an error of judgment? Ought he not to have deterred me?'

I was highly delighted with this explosion, and was half afraid that it would prove too much for our dear old tutor; but as the sounds of Dyver's voice died way, like the smoke after a volley, I saw him sitting calmly and quite unhurt by what seemed to me the murderous fire of his pupil.

'You mistake,' said he, 'the function of Logic. You can't blame Logic for not accomplishing what it never pretended or professed to accomplish. Logic is only responsible for the form, and not the matter of our thoughts. I shall explain form and matter in due time. Here it will be enough to say that it is when thoughts are self-contradictory or inconsistent that Logic interferes. If a man says, "The deep is shallow," Logic can correct him. If a man says, "All deep water drowns; this is shallow water; therefore it drowns," Logic can correct him, for the remark was self-contradictory, and the inference inconsistent. But if a man says, "This water is deep," Logic cannot interfere; and if a man says, "All shallow water is safe for bathing; this is shallow water; therefore it is safe for bathing;" and then, with Logic's consent, proceeds to bathe, and is drowned (for the water turned out to be deep), you couldn't blame Logic; for, as far as Logic was concerned, his thought was correct, his inference was logical; the only thing was that his data, his facts were not correct; and for these facts (which we call matter) Logic is not responsible. Given the facts, Logic can tell you whether the remark is self-contradictory or the inference inconsistent (*i.e.* can criticise the form). Every science must have its data or facts to start with, its materials to work upon. It's hard

upon the oarsman if we blame his rowing, when we
gave him a rotten oar to start with, and he loses the
race ; and upon the weaver, if we blame his weaving,
when we gave him bad wool to begin with, and the
cloth is useless.   And it is hard upon the logician,
when we give him faulty remarks to pass judgment
upon, or faulty facts or data (or premisses) to draw
inferences or conclusions from, and then blame him
for any mistakes that may occur from acting upon
his decision.   You bring this inference to the logician, "All speculation is profitable ; this Peruvian
affair is a speculation; therefore it is profitable."   "A
perfectly sound and correct inference ! " exclaims the
logician, and so you invest and lose your money ;
but it is you alone who are to blame, for you bring
him bad material to work upon.   Your remark,
"All speculation is profitable" is as faulty as the
most rotten of oars and the very worst of wool.'

The warmth of this reply quite restored my con-
fidence in our good tutor, and it was entertaining
too, as we knew he himself had lately lost some of
his money in the Peruvian Stocks ; and this is why
Dyver's illustration came home to him so well.

'After all, it cannot be denied that errors of form
in thought are more heinous than errors in matter.
We can make allowances for the latter, but not for
the former; and the science which remedies the
greater evil ought surely to be more highly valued.

'But to sum up briefly.   We have seen that

knowledge is of two kinds, particular and universal. Universal knowledge derived from particular is called science, and applied to particular, is called art. The name of the science or art depends upon the objects with which it is concerned. We have also seen that there are such things as particular correct thoughts, and that science and art can be introduced here, enunciating and applying the conditions or laws upon which those correct thoughts depend. And this is called Logic, and has only to do with the *form* as opposed to the *matter* of thought, expressions which we have reserved with the laws of thought for further explanation.'

CHAPTER IV.

FORM AND MATTER OF THOUGHT.

' BEFORE commencing our explanation of the form and matter of thought, I should like to answer any questions that may have occurred to either of you.'

' I am troubled about one thing,' said Dyver. ' What do we mean by saying a thing is "true" or "correct"—how would you express the meaning of these words?'

' There is no need whatever to resolve these words or further explain them. In Logic we start with particulars as in most, if not all, of the sciences (for some maintain that there are sciences which start from the universal and work downwards, as we shall find in course of time), and we must accept without question, or take for granted, these particulars, for whether we start from universals or particulars we must begin with assuming some thing, and believing it by instinct or intuition. For if we did not do so we could never start at all, nor have any basis for the construction of science, so to speak. In Euclid we start with certain postulates, and in botany we start with certain plants and trees, and in Logic we start with " correct and incorrect thoughts." If any-one asks us how we know a correct from an incorrect thought, we should reply by instinct or intuition, and that would be a sufficient answer.

' Still, to gratify your thirst for knowledge, I should say that a thing is true when it agrees with the impression of our own senses, and correct when it agrees with the decision of our own reason. For instance, if I heard a man say, " The comet is visible," and then saw it myself, I should declare the thing to be true; or if I heard a man say, " Comets are flying worlds," I should declare the thing to be correct enough, but I could not answer for its truth, whereas if he said " Comets are not comets," I should declare the thought to be incor-rect. Of course the impression upon the senses and the decision of the reason of our fellow-men is con-sidered almost as secure a pledge for the truth or correctness of a thing as that of our own. But this distinction I should not venture to insist upon.'

He then told us that to grasp the idea of form and matter we must do as we did with science, that is to say, we must first understand the idea generally, and then apply it to thought.

' Take any three statues; they may be of Mars, Apollo, and Venus, and all of gold, when they would differ in form and be the same in matter; or they

may be of gold, copper, and brass, and all of Apollo,
when they would differ in matter but be the same
in form.   And so with houses.   You may have a
villa, a mansion, or a public-house all built of brick,
or a villa of wood, brick, or stone.   The more homely
instance is still better.   A jelly, a cream, and a
blanc-mange may be all turned out of one shape or
mould, and then they would be different in matter
but of the same form ; and three jellies may be turned
out of three different shapes or moulds, and then
they would be the same in matter but different in
form.

    ' Apply this distinction to thought.   Any thoughts
may differ in matter and be the same in form, or
vice versâ, as the statues, houses, or jellies.   The
matter of a thought is that about which one is
thinking, and the form is the shape or mould into
which we cast the thought.   There are an infinite
number of things which go to make up the matter
of thought.   But the forms of thought are three (as
recognised by Logic) :—

    (1.) The Term or Concept.
    (2.) The Proposition or Judgment.
    (3.) The Syllogism or Inference.'

    I shuddered as I heard that word syllogism, so
often had I heard of it before, as one of those things
that nobody could understand,—and had we really

penetrated so far into the domains of Logic as to stir this giant from his lair? I was full of awe.

'Into one or other of these three forms or moulds Logic casts all the matter of thought; in other words, everything we know of or can conceive. Of the proposition or judgment, and the syllogism or inference, we have already spoken under the names by which they are known outside the pale of Logic, *viz.* remarks and inferences. To consider these three forms separately.

'(1.) The *Term* or Concept is, roughly speaking, the name of anything we can see or imagine. Horse, star, James, white, black, fire, philosopher, hippopotamus, priest, are all very different things to an observer who regards their matter, but to the Logician, who pays exclusive attention to the form, they are all terms or concepts. Fierce or tame, small or great, they are all equally terms to him. Were a Logician to be told first that a bedstead, and then that a rabid tiger were in the next room, separated by a thin partition, in so far as he was a Logician he would evince no emotion at the second piece of intelligence, but would simply reply, "Two terms," for the Logician is only concerned with the form.'

'In so far as he was a human being though,' said I, 'he would not stay long enough to say even that.'

'Quite so; because the matter of thought makes

a great difference in the regulation of conduct, but is not recognised at all in Logic.

'(2.) The *Proposition* or Judgment is a combination of two terms, and is the mould into which Logic casts all those " remarks " we spoke of above as the beginning of thought. " Horses are swift," " stars twinkle," " James is a haughty footman," " white is dazzling," "fire is comforting," "philosophers stoop," " hippopotami are large," and " priests are good," are all very different things to an observer who regards the matter of thought, but to the logician who pays exclusive attention to the form, they are all propositions or judgments. False or true, dogmatic or liberal, interesting or dry, they are all equally propositions or judgments to him. Tell him three facts—" The mixture of chloride of mercury with iodide of potassium produces a colourless liquid over a brilliant red precipitate," " beautiful women are fatal to the peace of man," and " rabid tigers infest the adjoining room," and in so far as he is a logician he will simply articulate " propositions," for it is the form and not the matter with which he is concerned.'

' But surely those long instances are more than the mere combination of a couple of terms,' said Dyver.

' It makes no difference how many *words* there are; there are only two ideas, though there are

D

several words; and this we shall explain when we discuss propositions more fully. To proceed :—

'(3.) The Syllogism or Inference couples together two propositions and produces a conclusion. Here, as in the case of propositions, it makes no difference whatever to the logician what the matter of thought may be. You may say to the logician " The mixture of chloride of mercury and iodide of potassium produces certain results ;

This is such a mixture,
Therefore this produces certain results ; "

or you may say, "Beautiful women are fatal to the peace of men;

My cousin is a beautiful woman,
Therefore she is fatal to the peace of men,"

and he will, in so far as he is a logician, say nothing more than "syllogisms," for such is the form or mould into which both these thoughts must fall. Thus is Logic said to be a formal science, and in this way is it concerned with the forms, ways, or modes in which people think.'

' I sincerely trust,' said I, laughing, ' that in so far as I become more of a logician my peace of mind will be less molested by what you call the "matter" of thought, for it would save much worry.'

That night, my nerves being, I suppose, in an

THE LOGICAL SAUSAGE MACHINE.

excited state from my attempts to follow my tutor's arguments—for application to study was quite novel to me—I dreamed a strange dream about the form and matter of thought. In the midst of a plain I beheld a huge machine. At first sight it resembled a coffee grinder, but on closer inspection it proved to be more like a monster sausage-machine. A college friend of mind was working it by a small handle, the perspiration pouring from his forehead.

On the left of the machine sat an old Professor, and on the right of the machine there was an indescribable confusion of all kinds of things ; there were birds, balloons, pyramids, and I could not tell how many more things there—a lion was just entering the machine, followed by a lady, a frog, a tortoise, and a clergyman. I should never have had my curiosity satisfied, had not the lion roared terribly in his reluctance to enter, thereby eliciting the following remarks from my friend at the wheel. ' Now then, in with you, you needn't make all that fuss ; you're not in the least degree formidable to us; why we had a whole menagerie through the other day, and trains, elephants, whales, worlds—all have to pass through this machine and become terms for the inspection of the great Professor Logic.'

For the first time I then noticed some things that looked like sausages issuing from the left side of the machine, only from having such big things

inside them they were swollen to the shape of eggs.
In one I observed a fish, in another a house, in a
third a bird, and in a fourth a man— all were painfully cramped for want of room.

'Ah! yes,' he went on, 'it's with our machine
here as it is with the common ordinary sausage
machine, never mind what the material is—so long
as the shape is right. Large or small, young or old,
flesh or stone, you must all pass through, for
Professor Logic isn't particular about the matter, all
he concerns himself with is the form,' and he laughed
as well as he could with his hard work, the thought
of the common ordinary sausage amused him so
much.

'Hard work, Frank,' said I.

'Indeed it is,' said he, without the smallest sign
of surprise at seeing me. 'Do you know the old boy
sitting there says, "One of the advantages, or practical utilities of being a student under me is that it
affords hard exercise to, and is therefore invigorating
to, your faculties," and I agree with him.'

'But what does he do with these terms when he's
made them?'

'I'll show you. I've done my share of turning
for to-day. The Professor likes us to work the terms
out *ourselves;* he says it makes us understand the
process, and I must get it up, you know, for my
exam. Still, I've done my share to-day.'

So, leaving the handle to another student, he

conducted me to the Professor's workshop, where I saw a vast number of terms from the machine linked together in couples, and they were called propositions. I afterwards learnt the meaning of the words written over them. There were also syllogisms or inferences composed of these same terms, and wondering at them, I awoke.

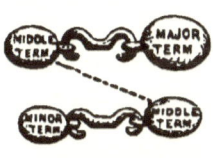

CHAPTER V.

THE RECONCILIATION, AND ALSO HOW LOGIC IS MORE
OF A SCIENCE THAN AN ART.

'WE can now proceed to reconcile the various definitions of Logic,' resumed Mr. Practical. 'According
to some, Logic is the science of reasoning. Science
is now no longer a difficult word to us, and reasoning
is the process employed in the syllogism (to speak
technically), or in the inference as opposed to the
remark (to speak roughly). As the syllogism implies
the existence of the proposition, reasoning is equivalent in this sense to thought; and, if we suppose
that science also implies art, the definition becomes—
Logic is the science and art of thought. It has also
been defined as " the science of the necessary forms
of thought," and so it has been called a "formal
science;" for, as we have seen, Logic is only concerned with the forms or modes in which people
think as opposed to the matter. It has also been
called "the art of thinking," *i.e.* the application of
universal laws to particular thoughts; and the
universal laws imply the existence of science. Lastly,
it has been well defined as "the science of the

conditions on which correct thoughts depend, and the art of attaining to correct and avoiding incorrect thoughts." '

'But which should we give as our definition?' I asked.

'Provided you were ready with an explanation, and thoroughly understood what you meant by the words, you might be content with this—Logic is the science and art of thought.'

'I begin now to realise the meaning of the expressions. "Science is knowing" or "a knowledge of what is," and "art gives rules for practice" and acts, and science is theoretical and art practical.'

'The next thing to be shown is that Logic is more of a science than an art; that is to say, more employed as a science than an art. Let us consider the results of the employment of Logic as a science and as an art, and then, if we find that we already have the one and not the other, we shall perceive that Logic is more useful as supplying us with what we have not, than with what we have. Roughly speaking, the results of Logic as a science are laws, and the results of Logic as an art are particular thoughts in accordance with those laws. Now we already have the particular correct thoughts. We do not want Logic to tell us when a thought is correct or not. The costermonger detects a contradiction or inconsistency in almost all cases as quickly as the Logician, though he may not be able to say why the thought is wrong. Whereas, we have not already the laws. Therefore Logic is more useful to us as a science than an art.'

Dyver looked a little puzzled, and then asked, 'If art were only manifested in production, or whether it were not also manifested in criticism?'

'I was about to add that in the case of criticism the art of Logic would be of great use; but, insomuch as production is the chief part of art, Logic is not held in good repute as an art. It is for this reason that the practical utility of Logic has so often been called into question. Of this we shall speak again. The case would be the same with the science and art of healthy breathing. It would be interesting and profitable to know the laws upon which healthy breathing depended, but few would have any idea of changing their manner of breathing by an application of those laws. It is thus that Logic is more of a science than an art.'

It so happened that we had our Logic lecture after supper this time, to enable us to make a sailing expedition in the daytime; and, partly from fatigue, partly from the effects of supper and work, I actually dreamt about Logic again. I thought I was wandering in the parks at Oxford, and I came across an old woman of so pitiable an aspect and such miserable attire that I could not forbear stopping to inquire what she did there and whence she came. She sat, with a basket on her lap, swaying slowly from side to side, and took no notice of me. At last she heaved a sigh and began, ' Ah, me! poor Dr. Logic (he ain't what once he was!). Good sir, I'm his errand girl. I takes round what he makes up. Deary me! day after day, door after door, the same

old answer. Fust of all I says, "D' ye please to require any of my master's beautiful works? I've two sorts here, I says; beautiful ' results of art' on this right side o' my basket." "What d'ye mean?" says they. "Why," says I, " pertikler kerrect thoughts—just made and ready for use—please to give 'em a look!" And they says, "O, then! we won't trouble you—we ain't *out* of them—we've got as many as we want of our own;" and 'ow can I ask 'em to buy what they've got aready? Ah, deary me! poor Dr. Logic; he ain't o' much account now-a-days. I catches 'em up, 'owever, and says, "Well, you ain't got these others in this left side—these splendid 'results of the science,' the laws upon which them pertikler kerrect thoughts as you've got a'ready dippends; and they says, "no, we ain't," and may be they'll take some o' them—that left side's all poor Doctor ever sells! But, bless yer 'art alive, sir; there's many of 'em as won't take even them. "We ain't got 'em? no! and we don't want 'em. None of yer humbuggy, cranky, theoretical stuff 'ere—practical results—that's what we want. Come—be off!"'

I woke, thoroughly moved to pity, and vowing I would never treat poor Dr. Logic so badly.

CHAPTER VI.

LOGIC THE SCIENCE OF SCIENCES AND ART OF ARTS.

'TELL us, Destrawney, what you mean by science of sciences.'

'It seems to me to mean "best of all sciences;" science—science—I forget the word—*par exemple?* no, no! something like that—dear me—'

'*Par excellence,*' suggested Dyver.

'That's it,' said I; 'science *par excellence.*'

'I should rather say that science of sciences and art of arts means the science which is concerned with every other science, and the art with every other art. And this may be put simply thus:—Thought is required for every science and art; for without thought we cannot ascend from particulars to universals, or descend from universals to particulars, and this is science and art. It is for this reason that the animals are destitute of science and art. They have sensation—that is to say, they can apprehend particulars, but general notions or universal propositions or inferences we have every reason to suppose they cannot attain to. The dog knows "this fire" and "this fire burns." But it is very improbable that the dog can form the general notion " fire," or the universal proposition "all fire burns," or the deliberate inference " all fire burns my nose ; master's cigar is fire ; therefore master's cigar burns my nose." Though the bird and the beaver and the bee evince marvellous sagacity, it is probably without any power to ascend to the universal ; or, at all events, to ascend consciously, as man does. Thought, then, is required for every science and art ; but thought itself is, so to speak, subject to its own science and art, Logic. Thought has its laws, and must obey them wherever it goes, if it would be called correct thought, as much as men have their laws and must obey them wherever they go if they would be called respectable men. Consequently Logic is the science and art of every science and art.

'To put it briefly. Where thought goes there the science and art of thought go also.

'But the science and art of thought were proved to be Logic.

'Therefore where thought goes, there Logic goes also.

'But thought goes into every science and art.

'Therefore Logic goes into every science and art.

'Therefore Logic is the science of sciences and art of arts.'

'Q.E.D.' muttered I mechanically and with an involuntary shudder at the thought that Logic seemed

SINDBAD THE SAILOR.

turning into mathematics; 'but still I don't understand yet.'

'Here is an illustration. The man with the high-forehead is Thought. Chained to his neck is Logic, holding perpetually before his eyes the rules he must obey, never mind what field he wanders through. Thought has a staff, for he is a great traveller. It would be impossible to enumerate the fields through which he passes. Four of them are given as specimens. Take botany. Thought lingers by a stream and gathers rushes. "Rushes are light," he remarks, and Logic is to all appearance asleep. He gathers more and murmurs, "Rushes are heavy." A smart blow on the head from Logic and a cracked voice crying, "Nonsense; WILL you look at your laws. How *can* they be both heavy and light? You meant when you had a big bundle they were heavy?—then I wish you'd say what you mean and not give me the trouble of showing you the law of contradiction." Thought meekly raises his head to read the laws once more, and catches sight of some stars in the sky. Gazing in wonder he meditates aloud, "Yonder's a lovely star; most assuredly that star must be a planet, for I know some stars are planets." A violent pinch from Logic and a heap of reproaches for the unwarrantable inference, and Thought wanders on patiently listening to the monitor that accompanies him wherever he goes, for though his voice is harsh, dry, and difficult to understand

Thought knows that the restriction is good for him, for he remembers that when, upon one occasion, he silenced that squeaky voice by diving so deep into metaphysics that the poor little monitor became insensible, he was caught in a whirlpool of circular arguments, dashed along wild torrents of fancy, and hurled into fathomless depths of conjecture to such an extent that it was with the greatest difficulty he ever escaped to dry land to become the laughing-stock of men. And many a time a Mrs. Cawdle has made such havoc of the distinction between universal and particular statements, by turning a deaf ear to the voice of Logic, that Thought has been exposed to sad ridicule and contempt. For upon Mr. Cawdle's *once* bringing in a friend late at night we find her indignantly demanding what he meant by day after day making a habit of bringing in mobs of the wildest of the wild to eat legs upon legs of cold pork and send girls miles for pickled walnuts in the middle of the night through depths of snow. Or, again, we find her violating the laws of contradiction by lecturing Mr. Cawdle for what he had *not* done, on the assumption that a man could both do a thing and not do it at the same place and time, *e.g.*, when she upbraids him at length for flirting at Greenwich fair, and upon his denying the charge retorts that if he didn't " it was no fault of his" and continues the lecture precisely as if he had pleaded guilty to the charge. Or, again, when she expresses herself as being very anxious to learn the Masonic secret, and at the same time declares it to be a matter of the utmost indifference to her; or dwells upon the pain caused her by some slight on the part of Mr. Cawdle, and at the same time entreats him not to allow himself for a moment to imagine that any conduct of *his* can possibly produce the smallest effect upon *her*; or, lastly, when she bitterly reviles him for the loan of 5*l.* to a friend, casting in his teeth at least 100*l.* worth of damages in the house that might have been repaired with that five pounds —thereby violating the law of contradiction to the extent of 95*l.* Consequently thought has no wish to rebel against Logic, whose sway is (in one sex at all events) undisputed.'

'I have read,' remarked Dyver, who seemed a little impatient of anything like homely instances —though I always liked them better than scientific ones—'that the words biology, zoology, chronology, &c., are equivalent to " Logic applied to life," " Logic applied to animals," " Logic applied to time, &c." '

' Quite so,' replied Mr. Practical, ' and it is always well to aid the memory as much as possible by understanding the derivation and meaning of the names you meet with.'

CHAPTER VII.

THE RELATION OF LOGIC TO LANGUAGE.

NEXT day Mr. Practical was compelled to visit the metropolis, and after a brief explanation of the relation of Logic to language he took his departure, expressing a wish that we should both endeavour to illustrate his meaning during his absence.

He told us that Logic was really connected with thought, but as it was impossible to get at thought (or at least at other men's thoughts) without the aid of language of some kind or other, Logic was so far connected with thought. The relation of Logic to thought he characterised as *primary*, while that of Logic to language was *secondary*. He also showed us how Logic had been held by some to be primarily connected with language, and so through errors arising from the inability of language to express exactly what we think, Logic had fallen into disrepute. The remedy for such mistakes was the 'knowledge of more than one language. As an instance of such errors he gave us the notion that the copula implied existence, and promised to explain this further. Dyver retired to his corner and I to

mine, and on the following day we produced a couple
of rough sketches.

This was Dyver's relation of logic to language.

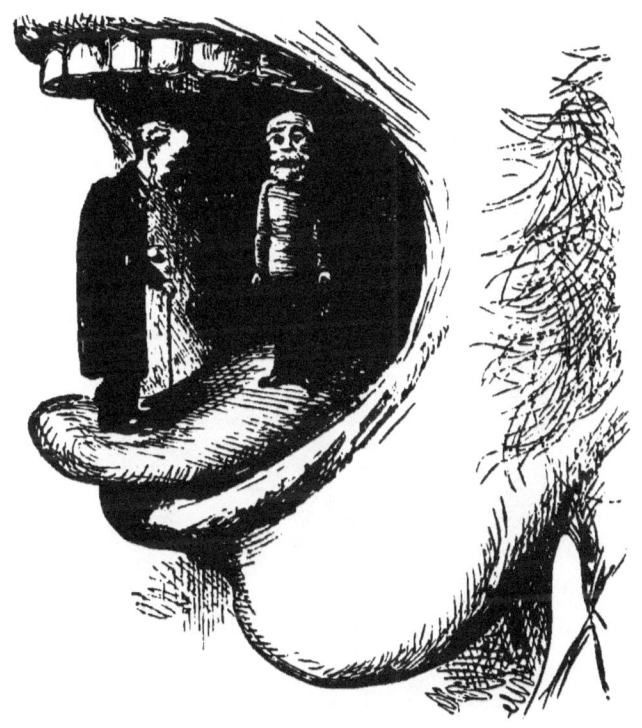

Logic pays a Visit.

Scene, a man's mouth. Thought peeping from
the throat. Language telling Logic (who has come
to visit Thought) that his orders are that no one can
see Thought, and that all communications must be
made through him (Language). 'So you'd better
by 'alf tell me what you want 'owever.'

And Logic mutters, ' What a coarse medium!

E

But there is no help for it.   Alas! what mistakes
and confusions will arise !'

And this was mine.

Logic in Trouble.

A friend of mine, named Jones, who lived in
lodgings, used to work with a tutor for Latin
prose. Jones was in the habit of leaving his productions daily at his tutor's, and calling next morning
to see the mistakes; but as Jones was not an early
riser his tutor would often step over to his lodgings,
and leave the corrected prose with the people of the

house, pointing out the faults, and begging them to point them out to Jones as soon as he appeared, for he was not yet to be seen. It so happened that the landlady's daughter was very attractive, and by some accident she always chanced to be at hand when Jones's tutor called. Scandal might have arisen from the protracted interviews in the course of which Jones's tutor pointed out to Jones's landlady's daughter the defects in Jones's Latin composition, had it not been well known that the *primary object* for which the tutor came was *not* to converse with the damsel, but to instruct the pupil, and being unable to see the pupil, he was compelled to employ the aid of a third person as a medium. His relation to the pupil was, so to speak, primary, his relation to the landlady's daughter secondary. After a time, however, Jones left, but the tutor still continued his visits, and ended by marrying the landlady's daughter. So Logic, led away by the allurements of language, has sometimes neglected thought, the object with which it is primarily concerned, and has thereby become involved in troubles and mistakes, and has fallen into bad repute.

Mr. Practical was highly pleased with both our illustrations, and added something to mine which I had not thought of. 'Had your unfortunate tutor,' he said, ' moved more in society, and known several of the fair sex, and become more inured to their attractions, he would not, in all probability, have

been led into a *mésalliance*, and in the same way
had Logic been conversant with *several* languages,
and become acquainted with their powers of mis-
leading and beguiling, instead of only knowing one,
Logic would not in all probability have been led into
entering upon a relationship with Language which
was likely to bring difficulties, error. and evil
repute.

CHAPTER VIII.

ALL THOUGHT IS COMPARISON.

' WE have seen,' resumed Mr. Practical, ' that Logic has for its subject-matter thought. What is thought? To give a rough answer to this difficult question, we must say that "all thought is comparison." We have already spoken of thought as consisting of remarks and inferences, for we may consider thought as unexpressed in the mind, or expressed in language. But these remarks are compounded of two things, something about which we are speaking, and something that we affirm or deny of it. If I say "dogs are animals" I am speaking about "dogs," and I say something about them—"that they are animals." Thought may therefore be divided into three parts—The term or concept; the proposition or judgment; and the inference or syllogism.'

'Do you mean by thought here the process of thinking or the results attained to?' asked Dyver.

'I should say the results attained to; of the faculties we shall speak hereafter, so that if we can show that the term, the proposition, and the inference or syllogism are the results of comparison, we shall have proved that all thinking is comparison; nor need you be alarmed, Destrawney, at any nice distinctions of the sort, for to pass your examination it is not necessary to notice them at all. Remember this :—

' 1. The *term* or concept is anything we can see or imagine. Our minds are stocked with concepts or terms formed in our earliest years. How came they there? Take an infant, hold a dog up to it, and the infant will recoil with a start of horror; it has experienced a mere undefined sensation of the presence of something strange, and therefore terrible. Repeat the operation daily, always taking care to accompany it with the word " bow-wow," and the child, from a comparison of the sensations, forms a concept, and associates these familiar sensations grouped into one idea with the name of " bow-wow," and if, after a time, you merely say " bow-wow," without introducing the dog, the child will manifest joy or terror according as it likes or dislikes the dog, proving that it has formed in its mind the concept or term dog, and can shut its eyes, so to speak, and see a dog though no dog is near. Thus, concepts are the results of the comparison of simple sensations and concepts expressed are terms. We gradually accumulate our concepts and distinguish one from another. As infants we had only a few under which to arrange all the objects that met our view. Papa, moo-cow, gee-gee, and bow-bow formed our stock,

then, and under one or other of these heads came
every man and every animal we saw. Nor can we
boast much now, for in botany, for instance, many of
us have one vague concept, " plant," under which
to arrange all the phenomena of that science, and in
geology it is generally considered a sufficient reply to
the question, " What is this?" if we say "A kind of
rock."

' (2.) As the term or concept is a result of the
comparison of simple sensations, so the proposition
or judgment is a result of the comparison of terms or
concepts, and henceforth we shall speak only of
terms and propositions, having shown that they are
identical with concepts and judgments. With propositions thought proper begins. It is true that
terms are necessary to form propositions, but they
do not by themselves constitute a thought. We
cannot have a brace of birds without single birds ;
but single birds do not by themselves constitute a
brace. Suppose our infant to have formed several
terms—horse, book, house, sun, large, dry, tall,
bright, &c.—and never to attempt to couple them
together, but simply to repeat them singly, we
should at once question its sanity, as being unable
to attain to a thought, for, as we have said,
man has an innate tendency to group together the
like, and this is the origin of thought. We should
exclaim impatiently, " Horse, horse, horse ! What
about it ? What is the use of going on repeating

house, house, house—sun, sun, sun? If you repeat them from your birth to your death you will not have expressed a thought!" Thought begins when the child, having formed in one part, so to speak, of his mind a term—for instance, book, derived from observations in his father's library, &c., and in another part another term, for instance, " nasty," derived from sensations of medicine, chastisement, &c., couples the two together, and exclaims "Books are nasty." Thus propositions are the results of a comparison of terms.

'(3.) Lastly, the syllogism is a result of the comparison of two propositions. When the child upon being told to open its spelling book says "nasty," and further, upon being asked why he thinks the book nasty, replies "all books are nasty," he has given utterance to a syllogism which in its full form would read "All books are nasty ; this is a book ; therefore it is nasty."

'We have traced, then, the gradual formation of the term from sensations, the proposition from terms, and the syllogism from propositions; and shown that they are all results of comparison; and these are the three parts of thought; they are also called (as we have seen) the forms of thought.

CHAPTER IX.

THE TERM.

'As thought,' resumed Mr. Practical, 'is divided into term, proposition, inference, so Logic, or the science of thought, is divided into three parts, under the heads of TERM, PROPOSITION, and INFERENCE; and you will find this analysis of the subject very useful.

I. *The term.*
    1. Definition of term and various kinds of terms.
    2. Connotation and denotation of terms.

II. *The proposition.*
    1. Definition of proposition and various kinds of propositions.
    2. The copula of a proposition.
    3. Distribution of terms in a proposition.
    4. Heads of predicables, or a list of the relations which the predicate of a proposition can bear to the subject.
    5. *Definition*, or propositions expressing the connotation of a term.
    6. *Division*, or propositions expressing the denotation of a term.

III. *The inference or syllogism.*[1]
    1. Definition of inference and various kinds of inference.
    2. Moods and figures.
    3. Principles, laws, and canons of syllogism.
    4. Reduction of syllogisms.
    5. Trains of syllogisms.
    6. Hypothetical syllogisms.
    7. Probable reasoning.
    8. The fallacies.

---

[1] The third division, 'inference,' includes inductive inference as well as deductive inference, or the syllogism. But we are only now concerned with that part of inference called 'syllogism.'

If you learn this analysis and can give some account
of each of the heads, your knowledge will be
sufficient for the purpose in view. The analysis is
taken from Mr. Fowler's " Deductive Logic," and I
refer you to that work or to Mr. Jevons's " Logic "
for information upon points that do not seem to
require further explanation.'

'I hope,' said I, 'you will not leave us too much
to our own reading, for somehow or other things
seem to me to be put so difficultly in books.'

'I will do my best,' he continued ; 'and first let
us discuss the term. It comes from " terminus," or
" boundary," because terms are the boundaries of propositions ; for a term is defined as anything that may
stand as the subject or predicate of a proposition.
If a term is to express anything we can see or
imagine, it is clear that we must give an exhaustive
account of all things if we wish to enumerate terms.
Now everything that we can see or think of must be
a thing or a quality of thing ; in other words, an
individual or an attribute of an individual. Mention
a few things, Destrawney.'

'A star, fair, whiteness, chair, Lexicon, beauty,'
said I ; 'fun, William, suicide.'

'Every one of these is either an individual or
thing, or an attribute or quality ; and you must content yourselves with these expressions, as the question
as to " What is the meaning of thing ? " is beyond
the sphere of Logic.'

Dyver here took occasion to observe that he knew that individual meant something incapable of further division, *i.e.* a simple whole; and that attributes meant the properties ascribed to such a whole, *e.g.*, tree and greenness, or man and reason, for you couldn't saw the tree or the man asunder without destroying them, and greenness and reason were said to be attributes of tree and man, and ' thing' and 'quality' meant the same as 'individual' and 'attribute.'

'Well, then, a term expresses either an individual or group of individuals, or an attribute or group of attributes; and this is as much as to say that terms are an exhaustive enumeration of all that we can see or think of. Take your fingers and thumb, and remember the five terms thus :—(See diagram on next page).

"If a term expresses an individual it is a singular term, *e.g.*, Socrates (thumb). If it expresses a group of individuals, it may either express the group and not each individual as well; or the group and each individual as well. Thus, the " BLACK WATCH " expresses a group of soldiers, but you can't call each soldier a " Black Watch ; " whereas " horse " expresses a group of animals, and you can call each of those animals a " horse." The former are called collective terms, *e.g.*, " the Black Watch " (1st finger); . the latter, common terms, *e.g.*, horse (2nd finger). If a term expresses an attribute or group of attributes, it is called either abstract (3rd finger) or attributive (4th finger), the abstract term being a substantive, the attributive an adjective.'

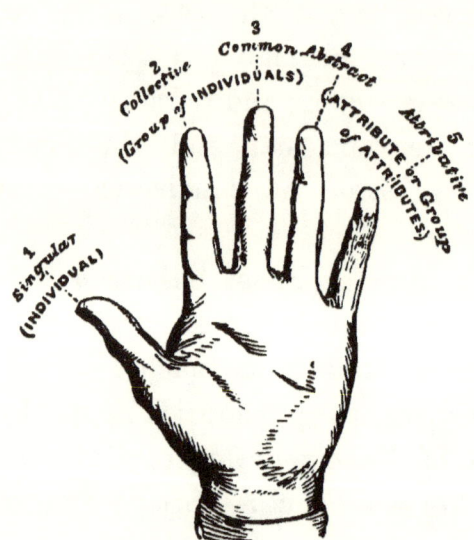

1. Socrates.
2. The Black Watch.
3. A horse.
4. Whiteness.
5. White.

N.B.—The second finger the largest, and the common term the most important.

The Logical Hand.

'Are there not several other kinds of terms?' asked Dyver.

'Yes; but it is my intention to work through our analysis first, and afterwards to explain briefly any names that may seem to interfere with the simplicity of our scheme of Logic.'

CHAPTER X.

CONNOTATION AND DENOTATION.

'Of the five terms, the common term is that with which we shall have most to do. Let us endeavour to explain connotation and denotation by taking a common term or class name as an instance. To the question "What is a man?" two answers may be given. Firstly, closing our eyes, we may give the attributes, the possession of which entitles a man to the name of a man—saying, "by man is meant the combination of the attributes of life, reason, &c.;" or secondly, walking to the window, we may point out Smith, Jones, &c. in the street and say, "Those are men." The first answer would give the connotation of man, the second the denotation. Do you follow?'

'No!' said I.

'Well, then, look at it thus:—Every common term, or class name, is a name given to certain individuals upon their complying with certain conditions, so to speak. You walk in the fields with a friend. You speak of trees, men, stones, birds; and your friend understands you to mean by them four distinct

things.   They all have attributes, and if those attributes were all the same, why should you not speak
of them all four as " trees ? " '

' But,' said I, ' their attributes are not the same ;
at least, I do not feel my senses affected in the same
way when I look at a stone and a tree.'

' Exactly so ! and these very attributes are what
we mean by the connotation. With the origin of
common terms we have nothing to do.   All we know
is that certain class names or common terms exist,
and that as fresh individuals present themselves we
enroll them under one or other of these names
according as they possess certain attributes, *e.g.*, class
man, attributes required for admittance rational and
animal qualities. Any individual possessing these
would be entitled to become a member of the class
" man," and so "rational and animal qualities " are
the connotation ; and Jones, Brown, Smith, Socrates,
&c., are the denotation of the common term " man."
Thus the connotation means the attributes in virtue
of the possession of which an individual belongs to
its class, and the denotation means the individuals
which, in virtue of the possession of certain attributes,
belong to a class. The connotation answers the
question " what ? " and the denotation answers the
question " which ? " *e.g.*, " What is a steam-ship ? " a
combination of the attributes of vessel with those of
steam ; in other words, " a vessel propelled by steam."
" Which are steam-ships ? "   The Great Eastern,

the Husband's boat, &c. ; or, again : last week when we went gull-shooting, I said, " Perhaps we shall get a puffin ; " and you asked, " What is a puffin? " and I replied a bird with such and such attributes, with form, colour, flight, &c. (connotation). Soon after several flew over and I said, " Look!—there—those are puffins " (denotation). The connotation of a term (under the name of *intension*) has been well defined as the qualities necessarily possessed by objects bearing the name, and the denotation (under the name of *extension*) as the objects to which the term may be applied. It has also been said that a term connotes attributes and denotes individuals, and the five terms are thus said to be connotative or denotative.

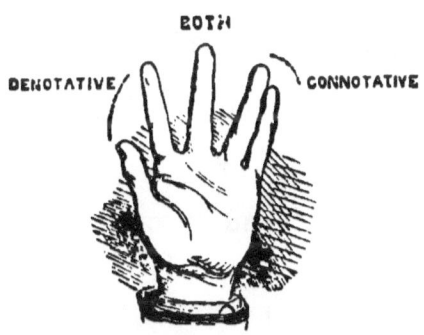

Denotative and Connotative Hand.

' The singular and collective have only denotation ; the abstract and attributive only connotation; and the common terms, both. For the singular term is a mere mark arbitrarily imposed upon an individual, *not* in virtue of the possession of certain attributes, but simply that we may be able to know it from its fellows. When our clergyman christened his boy "Harold" it was not because the boy manifested any peculiar attributes or qualities, but simply to distinguish him from his numerous brethren, so that (for instance) when they said "It was Harold who set fire to the barn," he might know which son to chastise. And what the singular term is to an individual, the collective is to a group of individuals—a mere mark affixed arbitrarily, for the sake of convenience.

'The abstract and attributive express only attributes, and therefore cannot be denotative; though both are capable of being regarded as common terms; for "redness" and "red" may be regarded as "red things," a class name or common term.'

'I can't quite understand the difference between singular and common terms with regard to connotation,' said I.

'Take an instance. In my stables I keep a mare known as "Fanny," and a cow known as "Polly." One day I say to my man, "John, we'll change these names; in future if men ask for their names, tell them the *mare* is called 'Polly,' and the *cow* 'Fanny.' "Werry good, sir," says John. Next day I add, "John, a further change is necessary; we will call the mare a cow and the cow a mare, and when men ask what they are, say this (the mare) is a cow and the other a mare." But

John stoutly refuses, and why? because the names "mare" and "cow" are common terms awarded to certain animals in virtue of their possession of certain attributes, whereas singular terms are mere marks arbitrarily affixed. Thus common terms are connotative, as implying the possession of certain attributes, and denotative as pointing to certain individuals, while singular terms are only denotative.

'You will see how the connotation of a common term increases as its denotation decreases, and *vice versâ*; for the more attributes required as a qualification for admission into a class, the fewer the individuals admitted.'

'Something like an examination which requires wide knowledge, or a club which is very exclusive,' I suggested, 'for the individuals who pass or are admitted are very few.'

'Quite so. Ask for a rose. The connotation is small: "a flower with certain peculiarities;" the denotation or individuals answering to that description very numerous. The common rose is plentiful. Then ask for a moss rose. The connotation is now larger. Candidates successful in the previous demand are now rejected as lacking the qualities implied by the word "moss;" and so the denotation is smaller, and you may continue the process until you get your connotation so large that there is only one individual, perhaps, that can satisfy all requirements —a prize plant and the property of a nobleman.

F

Take one more instance. You advertise in the
" Field,"—"Wanted, a servant; salary enormous.
A. B., High Street, Oxford." Next morning
the whole High, from Carfax to Magdalen, is a
seething mass of human beings. The Mayor is
furious, and compels you to withdraw your advertisement. After a while you insert the advertisement
again, only this time you write the word "male"
before servant. Once more the High is in an
uproar, and the Mayor interferes; but the numbers
are now reduced by about a half, though selection is
still hopeless. Angry with your own thoughtless-
ness, you determine to cut down still further the
number of applicants by adding more qualifications,
and eventually your advertisement reads, "Wanted,
a male servant, good looking, active, honest, with a
perfect knowledge of cooking and riding, who has
been in a training stable, and is able to translate
Livy and Virgil at sight, and to prepare all his
master's lecture-work." People now understand
why the salary is large, and upon looking from your
window you see two men, or one, or none, to answer
the advertisement. Thus the denotation dwindles
as the connotation grows.'

    ' Are not these contrary processes to be seen in
the growth of language,' asked Dyver, ' under the
names of generalization and specialization ? '

    ' Yes; and to remember which is which bear
in mind the fact that it is from the *denotation*

the names are taken. The word "paper" (or papyrus) originally meant "writing materials made of byblus;" it gradually came to mean "writing material made of rags, straw, or anything." The denotation has thus grown by the admission of several other kinds of writing material; and the connotation has diminished, for, at first, unless a candidate for the class "paper" (so to speak) could show that it possessed the attribute "made of byblus," it was not admitted; whereas now the class has been thrown open, and anything that possesses the attributes of "writing material" is admitted. On the other hand, "physician" originally meant φυσικὸs, "a man who studied nature." It gradually came to mean "a man who studied nature in respect of healing man." The denotation has here diminished, for the men who studied nature in other ways, *e.g.*, botanists, are excluded from the class; and the connotation has grown; for whereas, at first, any one who studied nature could have claimed admittance into the class physician, now more attributes are required to entitle individuals to admission into that class.'

That night I dreamed I was sitting in my rooms working, past midnight, when the door slowly opened, and a most strange figure entered. He seemed very unhappy. I asked his name and what he was? 'Ah, sir,' he sobbed, 'long years have made havoc of memory. Every one I meet I ask, What am I? and where is the class to which I belong? No one knows. A student like you I thought might give me help. My name? I've a faint recollection of being called " Scaly." ' ' Ah ! that,' said I, ' was a mere mark, a singular term.' ' A what ?' he gasped. ' The thing to do is to find your connotation and denotation; for you have attributes, and you're a connotative term.' ' Alas ! what mean these vile names ? ' ' Come with me,' said I, laughing, ' and we will find your class.' We went to a kind of Zoological Garden on a gigantic scale. The first cage we came to had a notice-board over the door. ' Class, Man—Connotation, Rational and Animal Qualities: none admitted unless they possess these qualities.' Innumerable swarms of individuals formed the denotation in that cage. ' Can you satisfy the requirement on that board ? ' said I. He said he could not, and we passed to the metal cage, and the plant cage, &c., but we could not find his class. At last a thought struck me. ' You are painfully thin,' said I; ' did you ever hear the name " Euclid ? " ' He started, and seemed violently agitated, muttering, ' That name, that name ! ' ' Come,' said I, as we hurried to that part of the gardens where the figures were. ' Do you feel very empty, as though you had nothing in you, and were only " lines enclosing a space ? " ' ' I do.' ' Then,' said I, ' you're a figure; we shall soon find your class now, only " figure " is

so vague and general; one might as well direct a
letter " Smith, America," as expect you to find
your class when you're only told you're a figure.
There are countless millions of them.' ' True,' said
he. We came to a sign-post just then, on which was
printed : ' To the Curvilineal Figures ; To the Rectilineal Figures,' with the hands pointing different
ways. ' Are you rectilineal ? ' I asked. ' Don't look
so scared ; it only means are your lines straight or
curved ? ' ' Straight,' said he, feeling his sides. We
hadn't gone far before we came to a place where the
road branched off into three directions, with a sign-
post marked, ' To the Rectilineal Figures of three
sides (Triangles) ; To the Rectilineal Figures of four
sides (Quadrilateral) ; and, To the Rectilineal Figures
of more than four sides (Polygon).' ' How many
sides have you ? ' I asked. ' Three.' ' Then you're
a triangle ! This is our way.' In spite of his over
flowing gratitude, we hurried on and found a large
building with the name ' Triangles ' over the en-
trance. The interior was divided into three com-
partments. For triangles with all sides equal
(equilateral) ; for triangles with two sides equal
(isosceles) ; and for triangles with no sides equal
(scalene). I had not time to answer to his cry of
delight, ' Oh, joy ! my scalene brethren ! ' before he
was in their midst, and I slipped away, and awoke
muttering, ' A scalene triangle, of course ! '

CHAPTER XI.

PROPOSITIONS.

'WE now pass to our second head, Propositions. A proposition compares two terms, asserting or denying one of the other. One of these terms is called the subject, the other the predicate; and the verb which intervenes is called the copula. To find the subject ask yourself, "What am I talking about—what is the *subject* of my remark?" To find the predicate ask yourself, "What do I say about this subject?" for prædĭco means "to assert."'

'Oughtn't the (i) to be long?' asked Dyver, 'and doesn't it come from *præ* and *dico*, to foretell.'

'No! there are two words; prædĭco, "to say before" (in the sense of time), and prædĭco, "to say before" (in the sense of place, *i.e.* publicly); and prædĭco is equivalent to κατηγορέω, and means "to assert, or declare;" hence predicate.

'Take an instance. "The sun is bright." Sun, subject—is, copula—bright, predicate.'

'Then in "Fair was her form," said I; 'fair, subject—was, copula—her form, predicate?'

'No! remember the questions you are to ask

yourself, and pay no regard to the position of the words. You are talking of " her form," and you say of it that it was "fair." "Her form" is subject, and " fair " predicate.

'The copula is some part of the present tense of the verb " to be." It is the sign of comparison, or of that innate tendency all men have to gather together similar things and separate dissimilar things. To state propositions logically, remember, they are ideas and not words, with which you have to do. A term often covers a multitude of words, e.g.,

Subject.	Copula.	Predicate.
The sun	is . .	bright.
Fires	are . .	burning.
The 'Faraday' . .	is not .	a-ship-that-will-convey the-telegraph-cable.
The-sound-of-the-waves-on-the-shore . .	is . .	beautiful.

There are three important remarks to make upon the copula.

1. It conveys no notion of *time*. If you want to express logically the fact " that Elizabeth was a good queen" say " Elizabeth—is—a-person-who-was-a-good-queen," and so with the future.

' 2. It conveys no notion of existence. The ancients said, "The unicorn is a non-existent animal." *Ergo*, "The unicorn exists a non-existent animal." *Ergo*, "The unicorn exists and does not exist," which is absurd. The confusion arose from the idea that " is " always means " exists," and is one of the mis-

takes we hinted at as due to the influence of language.
The copula only means " is equivalent to," or " not
equivalent to."

' (3.) It conveys no notion of probability, &c.
The question of the *modality* of the copula is the
question whether we are allowed to assert or deny the
predicate of the subject in a certain manner (*cum
modo*), *i.e.* conditionally, or with some qualifying
word joined to the copula, or whether we are only
allowed to assert or deny the predicate of the subject
simply.   The proposition, " Oysters are plentiful " is
called a pure proposition; " Oysters are possibly (or
probably, or certainly) plentiful " is a modal proposition.   Now, may the words " possibly, &c.," be
joined to the copula?   No.   All words of this description must be pressed into the subject or the
predicate, *i.e.* modal propositions must be reduced to
pure ones ; *e.g.*, "Prawns are possibly fierce " becomes
either   " That-prawns-are-fierce—is—possible," or
" Prawns—are—possibly-fierce."

' To impress upon your minds these three facts look
at this rough sketch.   The copula is the coupling
chain between two railway carriages, the subject
and the predicate, which form a train, and if Time,
Existence, or Probability wish to travel on the Logic
line they must get into one or other of these carriages.   It is unreasonable to suppose that they are to

be allowed to travel on the coupling chain—however full the carriages may be they must squeeze in somewhere, or not go at all.

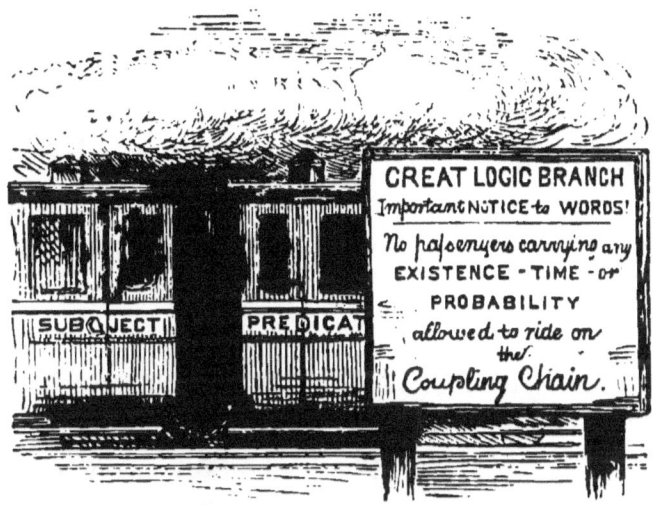

The Great Logic Branch.

'Lastly, propositions are divided into universal and particular according as their subjects are used in their full extent or not, and into affirmative and negative according as their copulæ affirm or deny the predicate of the subject. We may say, "All or no} Cretans are liars," or "Some Cretans are or are not} liars," and the first of these are universal propositions, the second particulars, and they are said to differ in *quantity* because they differ in the *amount* of Cretans to whom the term "liar" may be applied. Again, we may say, "Cretans are liars," or "Cretans are not liars," and these two are said to differ in quality.

We have thus four forms into one or other of which all propositions must fall before Logic can pass judgment upon them—

Universal affirmative .	" All Cretans are liars " .	. A
Universal negative .	" No Cretans are liars " .	. E
Particular affirmative .	" Some Cretans are liars "	. I
Particular negative .	" Some Cretans are not liars " .	O

A and I from AffIrmo, and E and O from nEgO. There are many other kinds of propositions which we shall explain afterwards, but these are the only forms with which Logic is generally concerned.'

'Under which of these four would you put, " Harold is brave," " The ' Black Watch' is a heroic band," " Virtue is rare ? " ' asked Dyver.

' Singular and collective terms rank as universals. When you say, " Socrates is bilious," you mean all of Socrates; and as for abstract terms like virtue, it is best to reduce them to common terms by saying, " Virtuous men are rare "; and as for common terms, remember Logic has no power to pass judgment upon propositions unless they are brought to one or other of the logical forms. The proposition, " Cretans are liars " is called indefinite, and with such Logic is not · concerned. You must specify the quantity or you can get no help from Logic.'

CHAPTER XII.

Mr. Practical began this lecture by shaking several blots from his pen on to a sheet of note-paper. 'This paper is wet with ink,' said he, 'and yet I can put my finger upon it without soiling my finger with ink.'

'Of course;' said I, 'you touch the paper between the blots; if the paper were wet all over with ink you could not do it.'

'It seems, then,' said he, 'that I used the words "this paper" in a partial sense, as meaning "part of this paper." Now terms thus used in Logic are said to be "undistributed," whereas terms used in their full extent are said to be "distributed." In the proposition, "All foxes are sly" the subject is distributed, *i.e.* we use "foxes" in its full extent; but the predicate is undistributed, *i.e.* the word "sly" or "sly thing" is not used so, for there are many other "sly things" besides foxes.

'To apply this distinction to our five terms, we find singular and collective terms are always distributed (*e.g.*, "Socrates is ill," where "All of Socrates"

is meant) and abstract and attributives may be treated
as common terms, and for common terms in propositions we have these two rules.

> (1.) All universal propositions distribute their
> subject.
>
> (2.) All negative propositions distribute their
> predicate.

' (1) Is obvious, for the sign of every universal proposition is "all" or "no," and "all" or "no" attached
to the subject of a proposition prove that we are using
that subject in its full extent. If we say "All men
are animals," or "No men are stars," we use the word
"men" in both propositions in its full extent.

' (2) All negatives distribute their predicate; for
in every negative proposition we may suppose ourselves to be excluding the things implied by the
subject from the class implied by the predicate. In
"No men are stars," "Some men are not poets," we
exclude the things "men" from the class "stars," and
the things "some men" from the class "poets." Now,
before we are justified in excluding anything from a
class we must look all through the class to be sure
that the thing does not belong to it. Suppose I say
"There are no blank pages in this book," which becomes "No—blank-pages—are—pages-in-this-book;"
without thoroughly looking through the "pages-in-
this-book" the result is an error, for one of you will
quickly turn to the fly-leaves to prove I am wrong.

Again, if I say, "There are no violets in this park," which becomes, " No—violets—are—flowers-in-this-park," I must look carefully through the class "flowers-in-this-park' before I am sure of this proposition ; in other words, I use the predicate in its full extent ; *i.e.* the predicate must be distributed. And so with the particular negative ; *e.g.*, " Some men are not poets," we must look all through the class " poets " before we can say that the " some men " referred to do not belong to that class.'

' I can't quite see how the particular negative distributes its predicate,' said I.

' Suppose I say " Some pheasants are not in the stubble field," which becomes " Some-pheasants—are not—things-in-the-stubble-field," I must look all through " things-in-the-stubble-field " before I can be sure of this remark, and even more carefully than if I had said " No pheasants are in the stubble field ; " for by " some pheasants " I might have meant " white pheasants ; " and though I found a thousand ordinary pheasants, it would not overthrow my remark, " Some pheasants are not in the stubble field ; " but in both cases I must look through the predicate. Hence it is far easier to affirm than to deny—to say " Some Irish girls are pretty," than " Some Irish girls are not pretty ; " for in the one case, if you met three instances, it would be enough ; in the other you would have to go through the whole class "pretty things" to be sure that your "some Irish girls" were not among them.

'Apply these two rules to A E I O, the four forms of propositions. A (universal affirmative) must obey rule (1) and distribute its subject; E (universal negative) must obey both rules (1) and (2) and distribute both its subject and predicate; I (particular affirmative) obeys neither and distributes neither; O (particular negative) must obey (2) and distribute its predicate.

'Remember this by the word A s E b I n O p (A subject, E both, I neither, O predicate).'

' A Word to the Wise.'

CHAPTER XIII.

HEADS OF PREDICABLES.

'THE heads of predicables is the name given to an exhaustive enumeration of the relations which the predicate of a proposition can bear to the subject. We are discussing propositions. In every one we find a subject and a predicate, but not always in the same relation to one another. Our aim is to find out all the different relations in which the predicate can stand to the subject. In "all men are animals" the predicate is a larger class including smaller classes. In "some men are yellow" the predicate is an attribute belonging to the subject. A deeper investigation than is necessary for us now has proved that these heads are five—genus, species, differentia, property, accident. The definitions of these may be thus expressed :—

A *genus* is a larger class, including smaller classes.

A *species* is one of the classes included in the genus.

A *differentia* is an attribute which is part of the connotation of a term, and marks off the species from the genus.

A *property* is an attribute which is not part of, but follows from the connotation of a term.

An *accident* is any attribute that is not a differentia or a property.

Don't look so scared, Douglas; remember we are only stating the ways in which a predicate can be related to its subject, and we are only concerned with common terms or class names. " What pudding was that we had yesterday ? "—" Dumplings."— "What kind of dumplings ? "—" Apple dumplings."— " Are there any other kinds of dumplings ? "—" Yes, currant dumplings, suet dumplings, &c."—" Very well; here dumpling is the genus, and apple dumpling, currant dumpling, &c.; the species, and the attributes " made of apple, made of suet, &c.," are the differentiæ; for they are that which makes one class of dumpling differ from another, and part of the connotation of the terms (the connotation being the meaning of the word).'

'What would be the property and accident?' asked Dyver.

' A property of " apple dumplings " would be that they are " good for the health," for though it is not a part of it, it follows from the meaning of the name " apple dumpling," for apples are good for the health.'

' But,' said I, laughing, ' you complained of indigestion after them.'

'True; but if " apple dumplings " are so made that they cease to be entitled to the name, you can't expect the properties which follow from the meaning of a term to remain when the meaning itself is gone. I refer of course to " apple dumplings " rightly so called. It was an " accident " of our candidates for that name that they were heavy ; for it is no part of, nor does it follow from the meaning of, apple dumplings that they should be " heavy." We have now all five heads, but be careful not to forget that they express *relationship between the two terms of a proposition.* If asked for the above heads, you would not say " dumpling," " apple dumpling," " made of apple," &c.; but you first give the connotation or meaning of the subject, and then give the various relations of the predicate thus :—

Given the term " This pudding " (which we will call T. P.).

(Connotation, or meaning—" apple dumpling.")

T. P. is a dumpling	(genus)
T. P. is an apple dumpling	(species)
T. P. is made-of-apple	(differentia)
T. P. is good-for-health	(property)
T. P. is heavy	(accident)

Of course the above example is for explanation only, and not for use. The instances you would use are the following pair, which are to be worked out in precisely the same way as the above :—

Given the term " Man."

(Connotation—" rational animal.")

Man is an animal	(genus)
Man is a rational animal	(species)
Man is rational	(differentia)
Man is able-to-do-Euclid	(property)
Man is unwell or well	(accident)

Given the term " triangle."
   (Connotation—"three-sided, rectilineal figure.")
  Triangles are rectilineal figures   (genus)
  Triangles are three-sided, rectilineal figures (species)
  Triangles are three-sided   (differentia)
  Triangles have two sides greater than the third (property)
  Triangles are large or small   (accident)

Notice the *arbitrary character* of these heads in two respects. 1. The heads themselves are not fixed, but vary according to your wish. 2. The connotation varies according to the point of view you choose to take.

' 1. The genus answers the question " What? " (τί; quid ?). The species, the question " What kind? " (ποῖον τί; quale quid ?). " What is a dog? " Answer, " An animal." Of course it is ; every one knows that ; that's only the genus. We want to know " what *kind* of animal? " We want the differentia, or attributes which distinguish " dog " from all the other classes in the same genus. We want the species, for the differentia plus the genus gives the species. The differentia (certain attributes, irrational, domestic, &c.) is added to the genus, and we get " animal with certain attributes, irrational, domestic, &c. " for the species. But what is species one moment may be genus the next. Man is a species of the genus animal, but a genus of the species European. There is, however, one genus which never stands as a species, and one species which never stands as a genus (the summum genus and the infima species). Man is a species of animal,

animal is a species of living being, living being of thing—and we can get no farther. So thing is the summum genus. So man may be subdivided till we come to a class which contains no classes, but only individuals, which is called " infima species." Even these are arbitrary, in so far as we may fix our own highest and lowest classes, provided we are consistent, as in the following tree of Porphyry. Man is re garded as an infima species.

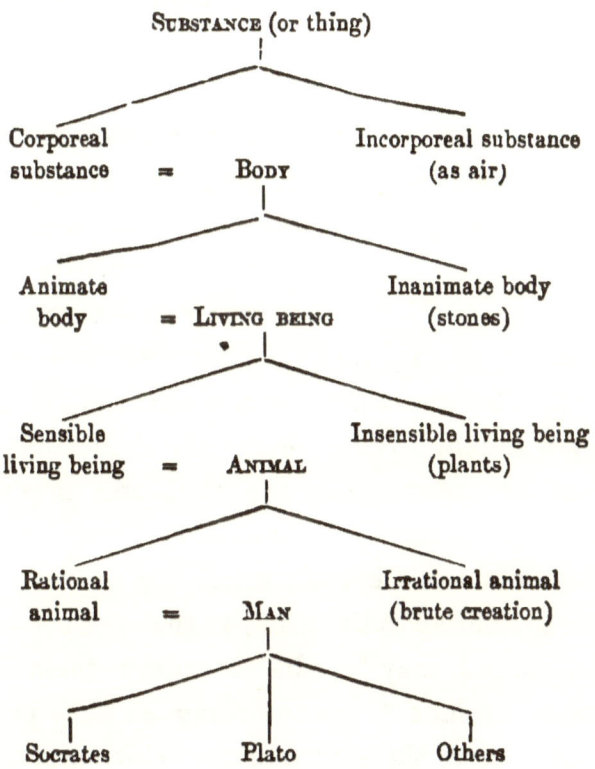

' Suppose you are being cross-examined thus :— " You speak of Plato. What is Plato ? "   " A man."

"What is a man?" "A rational animal" (genus "animal;" differentia, "rational"). "What the: do you mean by 'animal?'" "Living being with powers of feeling (*i.e.* sensible), for plants are 'living beings,' but they cannot feel; so 'sensible' is the differentia here, and 'living being' the genus." "But pray what do you mean by living being?" "A kind of body; the genus 'body' is divided into two species, 'body with life and body without life' (*e.g.*, stones and stocks). Body with life is 'living being.'" "Please tell us what body is? "Body is a kind or species of thing—things being divided into corporeal (as stones, &c.) or incorporeal (as air, spirit). Body is a solid, tangible thing." "What then is a thing or substance? " "A thing is a thing —there is no higher class; consequently no explanation or unfolding into simpler elements. It's a 'summum genus,' and I can't answer any more."

' 2. The connotation varies according to the point of view taken. We have already shown how connotation (or meaning) is the attributes implied by a name. I see certain individuals; I group them into a class and select certain attributes as characteristic of these similar individuals, so that if I see any more individuals I may be able to admit them into my class or exclude them according as they possess or do not possess these attributes. Obviously the attributes thus selected are not *all* the attributes possessed by the individuals (for it would be a hopeless task to enumerate them all), but only a few of them. Nevertheless, these few are called the connotation or meaning or essence of the name; and thus, the connotation of a word, which strictly should be all its attributes, is in reality only a few. The question is : " If all the attributes can't be taken, which shall be the privileged few to stand as connotation and to be applied as a test to all candidates for admittance into the class?" The answer is, "The attributes which are most prominent;" and prominence of course depends upon the side you stand on, or the point of view you take, and so there are as many connotations or meanings of a word as there are points of view from which to regard it.'

' Please give us an instance to make it clearer,' I gasped.

'Take the term " man." Here we have a group of individuals—a class. They resemble one another in countless attributes ; which of those attributes are we to select as connotation ? It's impossible to take them all. We must have some to apply as a test, and to exclude such animals as gorillas that clamour for admittance. Take the most prominent. "Rational and animal qualities " generally seem the most prominent. But change our point of view and our differentia (half-the-connotation) will change. Man and his attributes are like the round table we sit at ; the part that is prominent to you is not prominent to me. Here is an illustration.

'According to all these "his" connotation differs. The popular point of view seems the best. It is represented by ascending steps. First—things, the stocks and stones; then plants (life without feeling); then animals (life with feeling); and at the top men (life with feeling and reason); and here "animal and rational qualities" are the connotation, intension, meaning, or essence (for these all mean the same) of man. To make this still clearer, take the expression "a good country." This to the hunting man means "with good fields and fences," to the painter "with fine landscapes," to the botanist " with rare flowers," to the thirsty man " with frequent public-houses," and to the missionary, "with pious views." Here the connotation or essential attributes vary with the position of the spectator. Thus we hear people rebuke those who take mistaken views of life, " You seem to think life means nothing but 'eating and drinking,' " say they, where Logic would say, " You seem to think the connotation of life is eating and drinking qualities, and that 'man is an eating and drinking (instead of a rational) beast.' "

'Lastly, remember verbal (or explicative, or essential) propositions are those where the predicate *unfolds* the meaning or *essence* of the subject, and so tells you what you already knew, if you knew what the subject meant. Real (ampliative or acci- dental) propositions tell you something *more* than you necessarily knew, if you knew the meaning of the

THE ESSENCE OF MAN.

subject. Man is an animal, is verbal. Man is white or black, real. Of the five heads of predicables, the first three form verbal propositions, the other two real propositions.

' With regard to accidents, notice that they are of two kinds, separable and inseparable. " Some men are Europeans," is an inseparable accident. It is not a part of, nor does it follow from, the meaning of man ("rational animal") that he is a European, and so it is an accident; at the same time it is an abiding attribute, so to speak, as compared with such an attribute as " Some men are ill, or eating, &c.," which is called a separable accident. So we talk of men as "inseparables" when they are always found together, though they are not obliged to be together, and man and wife might be called " separables " because to-day they live together and to-morrow they may be divorced.'

CHAPTER XIV.

DEFINITION.

'You have now some idea of the signification of connotation, otherwise called intension, essence, or meaning. The proposition which expresses this connotation is called definition, as the proposition which expresses the denotation or extension is called division. It will be easy to follow now, if the idea of connotation is once grasped. Remember

1. What can't be defined.
2. All definition is incomplete in three ways.
3. All definition is relative.
4. All definition is "per genus et differentiam."
5. Rules of definition.

'1. Singular and collective terms can't be defined. Why? Definition gives connotation, and these two have none. You give your gardener wasps and ask him for the honey. He says they haven't any. So definition can't give you the connotation of singular and collective terms. Again, simple ideas can't be defined, for all definition is an unfolding, and you can't unfold what is already unfolded. So "thing or substance," the summum genus, can't be defined; and so sweet, bitter, harsh, &c. " What is love? " asks your unsusceptible friend. " Oh, it's very nice, but I can't explain it," you reply. " You must feel it to understand it; you'll know some day; it's one of those things that nobody can define. To understand what ' sweet' means, you must taste sugar."

' 2. *All definition is incomplete* because (*a*) the connotation it gives is only a part of the attributes As we have seen, " man " has an infinite number of attributes, but those implied by " rational animal " are considered enough to represent the class. (*b*) Even the attributes thus given are not in their simplest form—" rational " and " animal " are both capable of analysis (or breaking up into simpler elements). A being of another world may ask you, " What is man ? " and upon your replying " A rational animal," may go on " Yes, but what is ' rational,' and what is ' animal '? Are not these also open to further explanation? " and you would have to give a long explanation of both by help of Porphyry's tree. Thus a beggar at your door might receive a crust of bread, and depart grumbling at your charity as doubly incomplete—First, because of all your luxuries you only gave him a crust; and, secondly, because upon inspection even that crust turned out to be mouldy. So definition only gives us a few out of many attributes, and even those few are not in their simplest form. (*c*) All definition de-

pends upon the existing state of our knowledge.
Fresh discoveries upset old definitions. We talk of
plants and animals as two distinct kingdoms, but
the jelly fish looks very like a vegetable and certain
flowers strongly resemble animals. How if they were
discovered to be all one great kingdom instead of
two? All the definitions existing in the two kingdoms would be bundled away and vanish as the
coaches upon the invention of the rail.

'3. *All definition is relative.* Relative (re-fero)
means "has reference to something," and we have
seen how the connotation depends upon the point of
view you take. The prominent attributes, the "im-
portant-for-the-time-being" attributes are the connotation, and the prominence depends upon the position of the spectator. So definition is relative
—has reference to the point of view, and there are as
many definitions of a term as there are points of
view from which it can be regarded, *e.g.,* "good
country."

'4. *All definition is per genus et differentiam.* If
you are asked to define words in examination remember this. Always make your genus, find your
differentia, add the two, and you get genus + differentia = species = whole essence, connotation, intension, or meaning: and this is what definition gives,
for these four words mean the same. Ask yourself
first, "What is it?" and then "What kind of
it?"'

'But is it easy to get the genus and differentia?' asked Dyver. 'I saw the words botany, monarchy, money, metal, college, in a paper—how would you do them?'

Never having dreamt of attempting to define difficult words like these, I was astonished at the simplicity of the process as then given.

'Nothing easier,' he replied. 'As a last resource you can always call everything " a thing;" *e.g.*, a whip=a thing for driving; a ship=a thing for going on the sea, and so on ; but you generally know something more about the words, as here. Botany, quid? A science (genus). How different from other sciences? " Concerned with plants " (differentia); and hence the "quale quid," or species, "A science concerned with plants "=whole essence=connotation; *ergo* this is the definition. So monarchy. Genus?'

'Rule,' said Dyver.

'What kind of rule?'

'The rule of one.'

'So money=" a medium of exchange." Metal =" a substance hard and bright." College=" a society of men formed for the pursuit of knowledge.'"

'May I try " barrister?" ' said I.

'By all means.'

'A barrister is a man, genus (answer to what?), and his differentia is "pleads at law," and so the species," a man who pleads at law." Would this do?'

'Very well indeed.' You see now how *definition* gives the genus + differentia (=species.) *Description* gives you the properties and accidents, *e.g.*, "Man is able-to-do-Euclid," or " Man is a biped." '

'But if the connotation shifts with the position of the spectator, surely the properties and accidents from one point of view are differentiæ from another ? ' said Dyver.

'Quite so; before you begin to draw out your . genus, &c., you must determine your point of view. In the illustration above, the proposition "Man has a certain chest, posture, &c.," is an accident to the student who stands on the other side (so to speak), and to whom the " rational qualities" of man are the prominent ones; while to the student who regards man as " an animal of certain chest, posture, &c.," the fact that " man is rational " is an accident. Hence the definitions of one science become the descriptions of another, for the definition = genus + differentia, and the description = accidents or properties ; but the differentia becomes an accident or property if you change your science, *i.e.* your point of view, and so the definition becomes a description. Do you understand ? '

'Not clearly,' said I.

'You are in the train, sitting opposite a probable native of the country through which you are passing. You ask, "Is this good country here ? "

He looks at you carefully in order to get some idea of your occupation in life, that he may know the point of view from which you regard things. You look like a farmer. He replies, " Crops is fair." You have a white necktie on. " Ah, sir, there's a deal of drinkin'. " Perhaps you carry a hunting crop. He says, " Not much covert, sir"; or a microscope with roots in your pockets, he says, " Beautiful flowers about ;" or, lastly, you may be a weary traveller, and he will reply, " Poor accommodation, sir." But should your appearance give him no clue, he will reply, " Well, sir, that dippends upon what you mean by country. Yer see, what would make one man call a country good would be quite secondary like to another man. It ain't essential like to the man as looks at country from a 'untin' point of view that there should be purty landscapes, 'owever, &c." In other words, given a class name (*e.g.*, man) with an infinite number of attributes, the prominent attributes go to form the differentia, and all the rest of the round (so to speak) sinks into the secondary position of properties and accidents. Definition gives the former; description the latter. Change your point of view or science and your definitions become descriptions.

' 5. The rules of definition are not difficult to remember.

> (a) It must be *essential* (*i.e.* give the prominent attributes).

(*b*) It must be *adequate* (*i.e.* sufficient to distinguish the term from others).

(*c*) It must not be obscurum per obscurius, *i.e.* explaining an unintelligible term by terms if anything more unintelligible still; as if to enlighten a rustic you defined the soul "as a species of eutelechy . of a potential spiritual existence, my good man," or a flea thus: " A flea, madam, may be defined as an apterous hexapod."

(*d*) It must not be " circulus in definiendo." You must not in your definition come round to the term defined, and use it again. You must not say, " Metal is a metallic substance."

(*e*) No metaphors allowed. You must not define memory as " a *storehouse* of ideas." A metaphor is an image borrowed from one class of things and applied to another. When I speak of a lady " sailing " through a ball-room, the image is borrowed from ships. Metaphors are apt to mislead. From the lady's " sailing " in a room we might be led to imagine she could float in deep water, and the experiment might drown her; showing how dangerous metaphors are. So Logic does not admit them into definitions. Remember these rules

by the examples of their violation. " Man is a biped ;"
" Man is an animal ; " "the soul ; " " Metal ; "
" Memory," thus :—

CHAPTER XV.

DIVISION.

• DIVISION gives the denotation of a term as definition gives the connotation.   As there were five heads
under which to group all you need know about definition, so there are four heads here.   Only understand and you will find it easy to remember.

    1. Various kinds of division.
    2. Technical terms of division.
    3. Division and dichotomy distinguished.
    4. Rules of division.

'1. When you break a "plate" you divide it
into parts, but each part is not a complete plate;
whereas if you divide "plates" into soup plates, salad
plates, cheese plates, &c., each member of the division is called a plate still.   So if you divide "man"
into "arms, legs, &c." or into "Europeans, Africans,
&c."   The first kind of division is called partition or
physical division; the second, logical division.   One
divides wholes into parts, the other classes into
classes.   There is also metaphysical division or the
process of making abstract terms.'

'What are abstract terms?' asked Dyver.

'We have seen that the whole world around us

may be regarded as things and their qualities, or individuals and their attributes. As a matter of fact we never see things and qualities apart. Every quality we know of is always found in connection with the thing which manifests it. We never see " red " or " white " apart from " red thing," " white thing." But by help of imagination we can picture to ourselves " red " and " white " in the abstract. Now an attribute viewed in connection with its individual (con-cresco)[1] is said to be " concrete "; but viewed apart from its individual it is called "abstract " (abs traho), and the process of abstraction is called metaphysical division. So physical division divides wholes into parts, metaphysical separates individuals from attributes, and logical divides classes into classes.

' 2. The technical terms of division are the totum divisum, the membra dividentia, and the fundamentum divisionis. The totum divisum is the whole class to be divided. As in definition you take the species and resolve it into its component genus and differentia; so in division you take a genus, and by selecting some differentia or point of difference you find all the species that are included in the genus. The genus is then called the totum divisum; the point of difference, or the ground or source of the

<hr/>

[1] Concrescere = ' to grow with.' Abstrahere = ' to draw away.' So to speak, we draw away the green from the trees in bloom when we talk of ' greenness'—the green and the tree grew together.

division, is called fundamentum divisionis; and the species obtained the membra dividentia. Given the genus, "picture." Divide this. The question is "upon what principle? What is to be our fundamentum divisionis?" Take as F. D. (fundamentum divisionis) "manner of frame," and your M. D. (membra dividentia) become "pictures with gilt frames, pictures with wooden frames, pictures with all other frames;" or, take as F. D. "material" and your M. D. become "pictures in oil, pictures in water-colour, pictures in all other material." Loosely speaking, definition may be said to break up the species into the genus and differentia of which it is composed, while division, by the addition of differentia, builds up the species out of its genus and differentia.'

'Please explain this further,' said I.

'Given the common term, "guns." Divide this. You must select some point of difference if you want to find the species which form this genus "guns." Take as F. D. "manner of loading," and you get as your species or membra dividentia "breech-loading guns," "muzzle-loading guns;" or, again, F. D. number of barrels, and M. D. "double barrels" and "single barrels."

'T. D., houses; F. D., "material;" M. D., houses of brick, houses of stone, houses of all other material.

'T. D., horses; F. D., "use;" M. D., hunters, hacks, carriage horses, all other horses.

'As there are as many definitions, so there are as

many divisions as there are points of view, for each fresh point of view will start a fresh F. D. or source of difference from which to form differentiæ. The ordinary observer divides " animals " into " rational and irrational," the moralist into " animals with a conscience and without," the zoologist into " animals with such and such teeth, breast, or posture," the tailor into " animals to be clothed and not to be clothed," and so on. Each is determined in his choice of an attribute to stand as F. D. by the point of view he takes.

' 3. Dichotomy is a kind of division. Dichotomy (δίχα τέμνω) or the " cutting into two " always divides a genus into two species. It selects an F. D. and divides the class into one part that possesses an attribute implied and another part that does not, *e.g.*, men : F. D. " nation," M. D. " Asiatics and not-Asiatics." It depends upon the law of excluded middle. It is more useful in matters of which we know little. We should not divide the kings of England by dichotomy into Norman and not-Norman, and not-Norman into Plantagenet and not-Plantagenet, &c., because we know all the names of the houses. But in dividing mysterious things like the " corns in a horse's foot," if we divided thus—

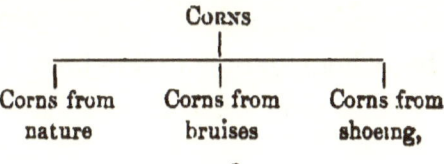

H 2

a corn might appear which was due to none of these causes, and we should have no class to enroll this new-comer in. Whereas, if we had divided thus,

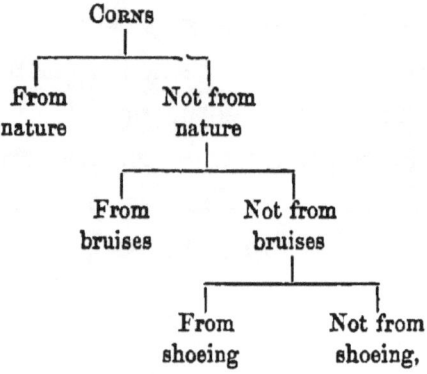

the new comer would be immediately enrolled in the class of "not from shoeing," and our classification would not have been found wanting.

'4. Learn by heart these rules :—

    (a) Each M. D. must be a common term.

    (b) The T. D. must be predicable of each M. D.

    (c) The M. D.s taken together must equal the T. D. or the division must be exhaustive.

    (b) There must be only one F. D.

'(b) Distinguishes logical from physical division. You can't say this piece of a plate is a plate, but you can say "soup-plates are plates."

'(c) The membra dividentia taken together must make up the whole genus divided, or the division will not be exhaustive, i.e. will not have exhausted the number of species in the genus. Divide the cartridges in your pouch according to colour—the membra dividentia are blue and green cartridges. After shooting away all the blue and green you find several yellow ones, and so your division was not exhaustive—luckily, or your shooting must have ended.

' (d) There must be only one F. D. or you have a cross division. You may have several F. D.s, but *only one at a time.* Thus—

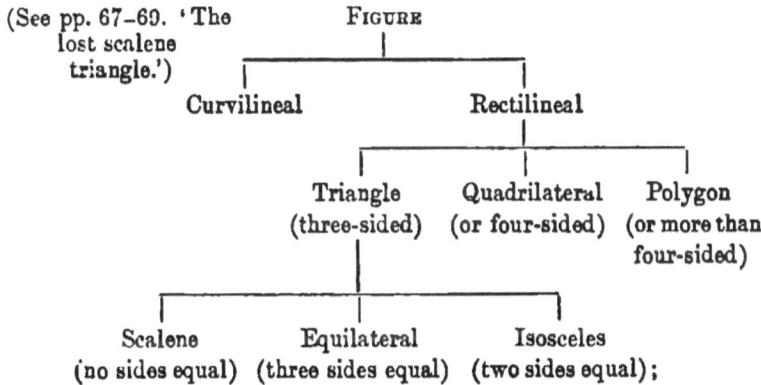

(See pp. 67–69. 'The lost scalene triangle.')

FIGURE

Curvilineal   Rectilineal

Triangle   Quadrilateral   Polygon
(three-sided)   (or four-sided)   (or more than four-sided)

Scalene   Equilateral   Isosceles
(no sides equal) (three sides equal) (two sides equal);

where subdivisions take place, and there are three different F. D.s in succession (nature of lines, number of sides, equality of sides), but only one at a time. A cross division would be to divide women into sisters of mercy, Americans, spinsters, and loud talkers, for in one division you have the F. D.'s " occupation," " nation," " marriage," and " manner of speech." Remember this :—

' " Your [1] M. D. is common, TeDious, exhaustive, but never cross," *i.e.*

(1) The M. D. must be common terms.

(2) Each M. D. must be a T. D.

(3) The division must be exhaustive (M. D. together = T.D.).

(4) No cross division (*i.e.* one F. D.).

---

[1] 'It has been said that this is equally true whether M. D. stands for Membra Dividentia, as above, or Membra Dividentes, *i.e.*, Doctors. But Logic is only concerned with the former and less violent rending.'

CHAPTER XVI.

INFERENCE.

'WE have now arrived at the third head, Syllogism. Syllogism is a kind of inference. Inference, from in and fero, means "the bringing in of new truth," or the "bringing in" of the same truth in a new form (for some say that in a syllogism we bring in no new truth). Now we may "infer" from particulars to universals, or from universals to particulars. From "John is mortal, James is mortal," &c., we may infer that "All men are mortal;" or from "No two straight lines can enclose a space," we may infer that these two ramrods cannot enclose a space. The first is called induction, the second deduction. Setting aside inductive Logic as beyond our limits, we subdivide deductive inference into immediate and mediate. Immediate—where we deduce one proposition *directly* from another, mediate—where we do so *not* directly, but by the help of a middle term. Hence inference is thus divided :—

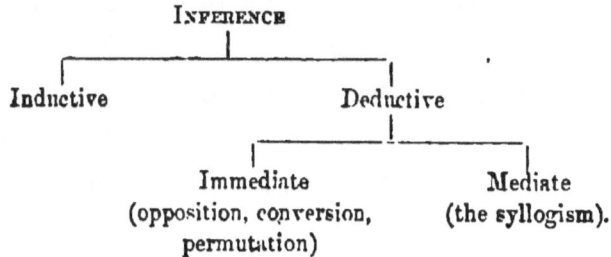

'To begin with immediate inference. (i) *Opposition*, where the truth or falsity of one proposition is inferred directly from the truth or falsity of another. The four forms *A E I O* are opposed in various ways, and this you will remember by this figure:—

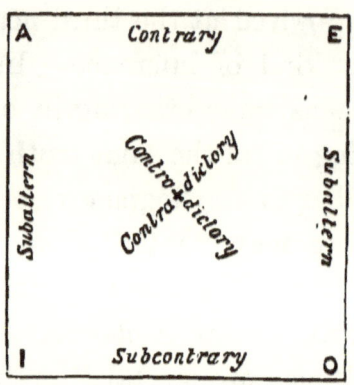

For the opposition of

A and *E* is called contrary opposition
I and *O* is called sub-contrary opposition
A and *I* or *E* and *O* is called subaltern opposition
A and *O* or *E* and *I* is called contradictory opposition.

The only case in which we can always infer the truth or falsity of one proposition from another is contradictory opposition, and you need remember nothing more than the names of the rest. To prove an adversary's statement false, use contradictory opposition. At first sight contrary opposition seems more powerful. If he says, for instance, "All Englishmen are black-bearded," you are strongly tempted to make a sweeping retort, "No Englishmen are black-bearded;" but if you do so, as a man who makes too eager a lunge, you expose yourself, for he replies "Jones, Smith, &c., *i.e.* some Englishmen, are so," and he produces them. In the first instance you should have replied, "Some Englishmen are not black-bearded," and this would have refuted his statement.'

'How can you call it an opposition between *I* and *O* (" Some men are maniacs"— " Some men are not maniacs "), and still more between *A* and *I* and *E* and *O* (" All men are mortal "—" Some men are mortal : " " No men are fishes "—" Some men are not fishes ")'? asked Dyver.

' There is an opposition here, but only a very mild one,' he replied.

' (ii) The inference called conversion makes the predicate the subject and the subject the predicate, and so gets a new proposition, *e.g.*, " Some men are bold leapers "—" Some bold leapers are men."

' This change is sometimes dangerous. From " All our loveliest companions are women " it would be a grievous error to infer that " All women are our loveliest companions." Or from " All dogs are animals that bite," that " All animals that bite are dogs," *i.e.* that there are no other "animals that bite" beside dogs. Here we must change the *quantity*, and say from " All men are animals " not " All animals are men " (for where would the bears be ?) ; but " Some animals are men.". Where no change of quantity takes place it is called "*simple conversion ;*" where a change takes place it is called "*conversio per accidens.*" *A* propositions are converted per accidens ; *E* simply ; *I* simply ; *O* not at all. There is one case, however, in which *A* converts simply—when the subject and predicate are of the same size, so to speak, or co-extensive. Thus, "All men are rational animals "— "All rational animals are men ;" but, as a rule, and unless you know that they are coextensive, convert *A* per accidens.'

'But if you are asked to convert things like, "Talents are often misused, &c.," ' said Dyver.

'It is easy enough, if you will only remember to bring them to one of your four logical forms, *A E I O*, before you attempt to convert them, and then you know whether they are converted simply or per accidens : *e.g.*,

" Fixed stars are self-luminous."

This proposition is not recognised by Logic except as an indefinite proposition. You must bring it to one of the recognised forms of Logic (*A E I O*). If you want it converted.

	Subject.	*Predicate.*	
All—	fixed-stars—are—	self-luminous.	(A prop. per accidens.)
Some—	self-luminous—are—	fixed-stars.	

i.e. Some self-luminous things are fixed stars.

" No one is always happy."

Logical form :—

No —	men—are—	always-happy.	(E prop. simply.)
No—	always-happy—are—	men.	

i.e. No things always happy are men.

"Some of the most valuable books are seldom read."

Logical form :—

Some—(of)-the-most-valuable-books–are–seldom-read. (I prop. simply.)

Some—seldom-read—are—the-most-valuable-books.

i.e. Some of the things seldom read are the most valuable books.

"Every mistake is not culpable."

Logical form :—Some—mistakes—are not—culpable. (O, not at all.)

For it does not follow that "Some culpable things are not mistakes;" nevertheless if you wish to convert an *O*, you can change it to an *I* by saying, "Some—mistakes—are—not-culpable"—and then it converts simply.'

'I understand these simple cases,' said I, 'but how would you treat "He jests at scars who never felt a wound"?'

'Be careful about your subjects and predicates. Remember the test questions, "What am I talking of?" "What do I say about it?" What am I talking of here? "He-who-never-felt-a-wound" —the subject; and of him I say that he "jests at scars." It becomes then in logical form;

All—who-never-felt-a-wound—are—people-who-jest-at-scars.

An *A* proposition, which is converted per accidens, thus :—

Some—people-who-jest-at-scars—are—those-who-never-felt-a-wound;

and this is obvious, for there are others who jest at scars (*e.g.*, hardy men) besides those who never felt a wound.

'Mistakes often arise from the predicate's position. Always ask yourself the test questions, and put your propositions into logical form before you attempt to convert them. Here is another. "No one is free who doth not command himself" = "No—men-who-do-not-command-themselves—are—free," an E proposition converted simply. "Life every man holds dear" = "All—life—is—a thing which every man holds dear," an A proposition per accidens. "Only the brave deserve the fair" = "No—not-brave, &c." "Nothing is beautiful except truth" = "No—not-truth, &c.," and "Every little makes a mickle" becomes when converted—"One of the things which make a mickle is every little."

'(iii) Lastly, permutation (or immediate inference by privative conception) infers one proposition from another by changing the quality alone, *e.g.,*

> All men are mortal,
> ∴ No men are immortal.
> No cowards are good soldiers
> ∴ All cowards are bad soldiers

> Some investments are made with risk,
> ∴ Some investments are not made without risk.
> Some investments are not made without risk,
> ∴ Some investments are made with risk.

And Mr. Jevons makes this very clear by diagrams, by help of which we might represent it thus :—

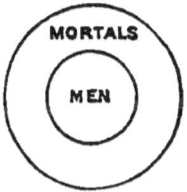

' Where it is plain that no part of the included class man can be outside the class mortals : .·. No men are immortal.

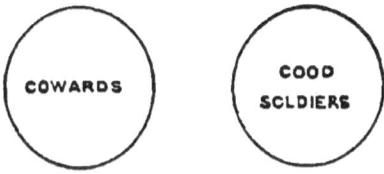

' No cowards are good soldiers ; they are two separate classes, and .·. All cowards are not good (*i.e.* bad) soldiers.

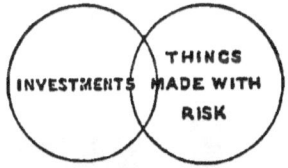

' Here the two classes overlap one another; and, as some of the investments are things made with risk, it is clear that the part of investments which overlaps is not without the other circle, *i.e.* is not without risk, and the last case is clear by the same figure.

' N.B. These diagrams also serve to illustrate the two kinds of conversion. The first proves that from

" All men are mortals " you get " Some mortals are men ; " the second, that from " No cowards are good soldiers " you get " No good soldiers are cowards ; " and the third, that if " Some investments are things with risk," " Some things with risk are invest- ments ; " and that if " Some investments are not things with risk " you can infer nothing, for it does not follow that " Some things with risk are not in- vestments," for we happen to know that some things with risk *are* investments.'

That night I had one of my Logic dreams. I thought I was walking through a fair. A man was shouting from a platform, ' This way for the real, live, strugglin' propositions ! Hevery kind of hopo- sition 'twixt them as differ in quantity or quality, or both ! ' I entered and saw all kinds of fights going on between things with small waists, like wasps.

The Fight of Propositions.

The spectators seemed to take very little interest in the fighting, except when, as every now and then it happened, a small animal engaged a large one and threw him and mortally wounded him. Then they shouted with delight. For the other combatants made a very poor fight of it. The man told me that what made them fight was that they differed in quantity or quality, or both. Suddenly, I heard shouts of 'Contradictory, contradictory!' and, hurrying to the spot, saw one of these mortal fights. The smaller animal had 'Particular' branded on its back, the larger 'Universal.' The spectators applauded loudly.

Death of the Universal.

Quitting this tent, I heard another man crying, 'Come and see the operations!' I entered a kind of hospital ward. They were operating upon animals similar to those I had seen fighting. They kept cutting off their heads and their bodies, and sticking them on again; the heads where the bodies were and the bodies where the heads were. 'They don't seem to feel it in the least,' I remarked to a bystander. 'Ah! said he, they be E's and I's; wait till you see an A done.' Very soon an A did come, and they had a severe struggle to convert him (for they called the operation 'conversion'), and he lost so much blood in the operation, that they all cried, "What a quantity lost! Oh! what a quantity!" Certainly he looked much thinner after it, as if he had gone through an accident (per accidens).   Last came an animal that kept struggling so violently that the operator said, ' I hope I don't hurt you? ' and he replied, ' O, not at all; ' at which the operator turned pale and let him go at once.

After that they experimented on one of these animals to show that the addition of a negative to each end of them made no difference to them.   They hung one up by his waist, like a pair of scales, and fastened something of equal weight to his head and his body, so that the balance remained as before. ' There,' said the operator, ' that's permutation— change of quality alone.'   At these words several of his audience looked puzzled; but when he added ' otherwise denominated immediate inference by privative conception ' there was a wild rush to the door, and I was carried along with the panic-stricken crowd, and woke, repeating to myself, ' Opposition, quantity or quality, or both, conversion, subject and predicate, permutation, quality alone, and by this I remember them always.'

CHAPTER XVII.

SYLLOGISM.

'MEDIATE deductive inference, or syllogism, is the assertion of a third proposition in virtue of two other propositions. Some say that only inductive inference is entitled to the name of inference. We shall regard the syllogism as an inference for reasons we will afterwards state. Take the instance,

> " All men are mortal,
> Socrates is a man,
> ∴ Socrates is mortal."

'Here there are three terms. Socrates, man, mortal. The term "man" occurs twice, for it is the middle term through which Socrates and mortal are compared. Hence the name "mediate inference."

'Now there are three things to remember about the syllogism.

(A) The *principle* upon which it depends.

(B) The *rules* which test its validity.

(C) The *canons* of its figures.

'(A) The principle upon which all syllogisms depend is that " things which are equal to the same are

I

equal to one another." In the above instance the
conclusion is "Socrates is mortal." Our object at
starting was to establish a comparison between
"Socrates" and "mortal." But it so happened that
we could not bring those two terms directly together, and we employed the help of a third term.
You have a station at the foot of the Alps on the
north side, and another at the foot of the Alps on the
south side. You can't make a tunnel. How do you
effect a junction? You must ascend to the summit,
and from the summit you must descend. So to effect a
junction between the terms "Socrates" and "mortal"
you must take "man" as a middle term. Thus:—

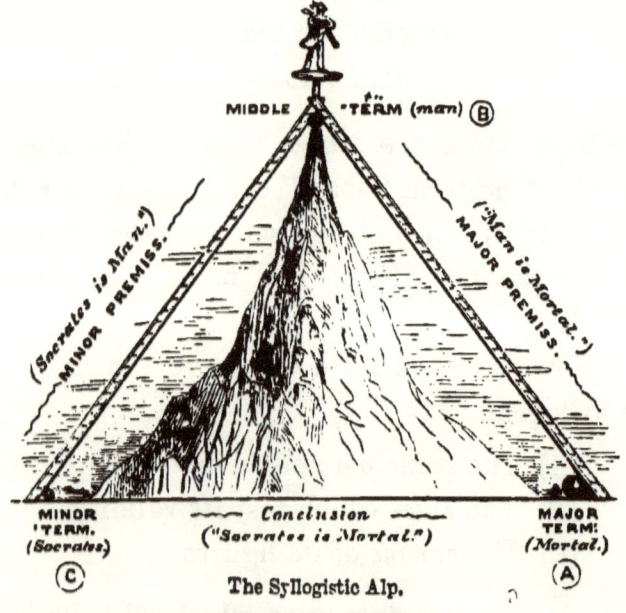

The Syllogistic Alp.

Here we are asked to establish a comparison
between "Socrates" and "mortal." Unable to see

through an intervening hill, we climb by ladder to the point man, and then we can see both " Socrates " and " mortal." From the conclusion, remember, we get our names. The predicate of the conclusion is the major term (for the predicate of a proposition is generally a larger class including the subject, as " mortal " is larger than Socrates), and the subject of the conclusion is called the minor term. Hence, major and minor premiss. The proposition which connects the major term and middle term is called the major premiss, and that which joins the minor and middle terms is called the minor premiss. The former always comes first. Unless the principle that " things which are equal to the same are equal to one another " were true, it would not follow that " Socrates " and " mortal," which are equal to the same (" man "), would be equal to one another.

' (B) The rules of the syllogism. We now know that a syllogism consists of three propositions, containing amongst them three terms : minor, major, and middle. Out of the propositions A E I O we can frame a large number of such triplets, e.g., " All animals can feel, all cats are animals, therefore all cats can feel," which would be all A propositions, or A A A; "No angels are human, all women are human, therefore no women are angels," which would be E A E, &c.'

' But would they all be true or valid ? ' I asked.

'That we shall find out by the application of

rules. First of all, Douglas, look carefully at the two syllogisms I have given you, and tell me if you see anything wherein they differ from one another.'

'To be sure,' said I; 'one is about cats and the other about angels.'

'True; but that is a difference of matter. What I want is a difference of form, and lest you should be tempted by the flesh of the propositions, so to speak, let us strip them of all but their bare bones; let us employ ciphers that in Logic we may not be misled by the matter. Always use the same ciphers. A for[1] the major term, B the middle, and C the minor. Thus:—

A. All animals can feel		All B is A	(where B = animals,
A. All cats are animals	= (i.)	All C is B	„ C = cats,
A. Therefore all cats can feel		∴ All C is A	„ A = things that feel.)

E. No angels are human		No A is B	(where B = human.
A. All women are human	= (ii.)	All C is B	„ C = women,
E. Therefore no women are angels		∴ No C is A	„ A = angels.)

Now do you see any difference between these two cipher syllogisms?'

'Yes,' said I; 'the propositions that compose them are different. In the first you have "all," "all," "all," and in the second "no," "all," "no."'

'Precisely so, and syllogisms differing thus are said to differ in *mood*. The syllogism A A A differs in *mood* from E A E, and so moods mean the various arrangements of propositions in syllogisms. But there is yet another difference.'

---

[1] Do not confuse this A (MAJOR TERM) with Proposition A—the first of the four A E I O.

Neither of us could detect this until we were told to mark the position of B, or the middle term in the premisses, and then we saw that B was subject in the major premiss (top proposition) of syllogism (i), but predicate of major premiss of syllogism (ii).

'Here, then, is another point in which these triplets of propositions, called syllogisms, may differ. And as syllogisms differ in *mood* according to the arrangement of their propositions, so they differ in *figure* according to the position of the middle term in those propositions. There can only be four figures. Where the middle term is subject in the major and predicate in the minor premiss, where it is predicate in both, where it is subject in both, and where it is predicate and subject; *e.g.*, take the mood A A A in all the four figures :—

A. All B is A		
A. All C is B	} Fig. I.	
A. ∴ All C is A		
A. All A is B		
A. All C is B	} Fig. II.	
A. ∴ All C is A		

A. All B is A		
A. All B is C	} Fig. III.	
A. ∴ All C is A		
A. All A is B		
A. All B is C	} Fig. IV.	
A. ∴ All C is A		

We know there can be only four figures; the question is, how many moods can we have in those four figures. The four figures may be remembered by the front of a collar.[1] As to moods, it is clear that out of the four forms or propositions A E I O a large number of different triplets can be made. We

---

[1] See next page. The figures are thus easily remembered; \||⁄, these lines being taken from the position of the middle term as marked above. For further remarks upon the figures see Appendix A (i.).

may have A A A in all the four figures, and A A E, A E E, and so on; these will be all syllogisms, *i.e.* triplets of propositions comparing two terms through

The Front of a Collar.

a third; but whether they will be all *valid* syllogisms is another question.   Supposing we act upon the conclusion we get from a syllogism and fall into error, we should at once turn upon Logic and abuse it.   But Logic does not warrant as valid every combination of the propositions A E I O in syllogistic triplets.'

' I don't quite understand,' said I.

' Take the mood A A A in the second figure, and tell me whether its conclusion is true :—

A.   all A is B
(fig. 2) A.   all C is B
A. ∴ all C is A.'

' It seems right, as far as I can judge,' said I.

' By this syllogism a man might pray for a dissolution of his marriage on the ground that his wife was married to a crocodile; for

" All crocodiles are animals,

All men are animals,

Therefore all men are crocodiles." '

'In this shape it certainly sounds wrong,' said I.

'Of course; and there are many other of the possible moods which will be found to be not valid in the same way. A mood not valid in one figure may be valid in another. There are sixty-four possible moods. How many of them can we admit into the four figures as valid syllogisms? That all the moods are not valid in all the figures is clear from the above instance, where the mood A A A in the second figure is found to produce an unheard-of conclusion. By what test shall we try pretenders to the name of valid syllogisms? The answer is, by applying certain rules to them; and if they do not violate these rules, they are valid; and if they do, they are not. These rules are eight. Do not be alarmed; they are all eight wrapped up in four lines easily learnt. To explain these rules :—

'(1) The *middle term must be distributed* at least once. For unless it be used in its full extent once, it may be used first for one part of itself, and secondly for another part of itself; *e.g.*

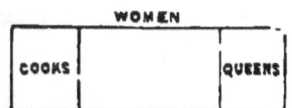

This whole figure represents the class "women;" parts of this class are "cooks" and "queens." Hence : —

All queens are women,
All female cooks are women,
∴ all female cooks are queens.

Had the middle term " women " been once used in its full extent, this false conclusion could not have been drawn.

' (2) *You must only have three terms,* otherwise the *principle* of the syllogism is violated, "things which are equal to the *same,*" &c. An ambiguous middle term is the same as two terms, thus :—

All chests are boxes,
Part of me is a chest,
∴ part of me is a box,

where " chest " means " box " in the major premiss, and " breast " in the minor.

' (3) *Two negative premisses prove nothing.* For if we wish to compare two things through a third, and neither of these two things is connected with the third, we can draw no conclusion. From " Socrates is *not* an elephant," and "Pugilists are *not* elephants," I should be in no way helped as to the question whether Socrates was a pugilist or not.

' (4) *Two particular premisses prove nothing.* For they would be O, O, which violates (3), or I, I, which violates (1), or I, O, which will be found to violate (5).[1]

' (5) *If either premiss be negative, the conclusion must be negative.* For if, after taking our middle

---

[1] For further explanation see Appendix A (ii.).

term as a medium of comparison, we say that one of the two things to be compared is equal to the middle term and the other is not, it follows that the two things to be compared cannot be equal to one another. If we want to compare " Socrates " and " stone " through the medium " animal," we say " Socrates is animal, stone is *not* animal," and the conclusion must be negative, " ∴ Socrates is *not* stone."

' (6) *If either premiss be particular, the conclusion*[1] *must be particular.* I will prove this rule afterwards.[1]

' (7) *Let no term be distributed in the conclusion, unless it has been already distributed in the premisses.* Otherwise we argue from part to whole. Violation of this rule, if it be in the case of the minor term, is called " illicit process of the minor," or, if it be in the case of the major term, " illicit process of the major."

' (8) *Let not the conclusion be negative, unless one*[1] *of the premisses is negative.* For we could not say, " Socrates is not stone " as a conclusion, unless we had a " not " in one of our premisses; *e.g.*, " Stone is *not* animal," " Socrates is animal."[1]

' Further explanation of these rules I do not think to be necessary, but you can find it in the books of Mr. Jevons or Mr. Fowler. For our purpose it will be enough to remember them by heart, and this you do by learning off these four hexameter lines :—

---

[1] For further explanation see Appendix A (ii.).

Distribuas medium, nec quartus terminus adsit,   (= Rules 1 and 2.)
Utraque nec præmissa negans, nec particularis ;   (= Rules 3 and 4.)
Sectetur partem conclusio deteriorem,   (= Rules 5 and 6.)
Et non distribuat, *nisi cum præmissa*, negetve.[1]   (= Rules 7 and 8.)
  (=unless when the premiss does so.)

'How do you translate the third line?' asked
Dyver.

' " Let the conclusion follow the weaker part."
The negative is thought " weaker " than the affirmative, and the particular than the universal. Thus
this line conveys rules (5) and (6). And before pro-
ceeding any further I must beg of you to learn these
four lines by heart, in order that you may move
amongst the syllogisms, armed, so to speak, with a
test, whereby you may prove their validity, and
dispel the false ones as evil spirits are dispelled by a
charm.'

[1] Thus rendered by some young lady friends :—
  'Distribute the middle, and have only three ;
  From two part. or neg. premisses proof cannot be ;
  Conclusion with part. or neg. premiss must co,
  And be dist. or neg. only if premiss be so.'

CHAPTER XVIII.

SYLLOGISM.

' WE are now ready to encounter all moods in all figures ; and, by the help of our "Distribuas medium," &c., to admit the true, and reject the false. Take A A A in all four figures :—

Fig. i.

A. All B is A
A. All C is B
A. ∴ All C is A

[Violates no rule. Call it "bArbArA."]

Fig. ii.

All A is B
All C is B
∴ All C is A

[Violates "Distribuas medium" —for A props. do not distribute their predicate. Remember ASEBINOP. Call this " undistributed middle."]

Fig. iii.

A. All B is A
A. All B is C
A. ∴ All C is A

[Violates " Et non distribuat nisi cum præmissa." C is distributed in conclusion but not in premiss (ASEBINOP). Call this "illicit process of the minor."]

Fig. iv.

All A is B
All B is C
∴ All C is A

[Illicit process of the minor, as in fig. iii. Remember ASEBINOP.]

N.B.—For ASEBINOP refer back to the "Distribution of terms in propositions," pages 75-78.

Take E A E in all four figures :—

Fig. i.

E. No B is A
A. All C is B
E. ∴ No C is A
[Violates no rule. Call it
"cElArEnt."]

Fig. ii.

No A is B
All C is B
∴ No C is A
[Violates no rule. Call it
"cEsArE."]

Fig. iii.

E. No B is A
A. All B is C
E. ∴ No C is A
[Illicit process of the minor.
From *part* of C in premiss we
argue to *whole* of C in con-
clusion."]

Fig. iv.

No A is B
All B is C
∴ No C is A
[Illicit process of the minor.]

Thus, amongst these eight possible moods, we have
found three valid ones—Barbara, Celarent, and Cesare;
and by going through all the possible moods (which
you should do for practice), we should find the following valid ones :—

Barbara, Celarent, Darii, Ferioque, *prioris* :   (*i.e.* in fig. i.)
Cesare, Camestres, Festino, Baroko, *secundæ* :   (*i.e.* in fig. ii.)
Tertia, Darapti, Disamis, Datisi, Felapton   (*i.e.* in fig. iii.)
Bokardo, Ferison, habet : *quarta* insuper addit
Bramantip, Camenes, Dimaris, Fesapo, Fresison (*i.e.* in fig. iv.)

These hexameter lines should be learnt by heart, as
there is much contained in them.'

'We have not yet seen what I have much desired
to see,' said I—'an illicit process of the major.'

'A. All B is A   All herrings are fishes
E. No B is C   or   No herrings are mackerel
E. ∴ No C is A   ∴ No mackerel are fishes;

and if ever you have to examine arguments, put them
first into ciphers, if possible, in fig. i.; and then, by

the help of "Distribuas," &c. and "ASEBINOP," you
will quickly find their weak point. Thus, in the
instance above, it is hard to find the flaw in the
herring-shape, but in the cipher-shape ASEBINOP
finds A undistributed in the major premiss, and distributed in the conclusion; *i.e.* an argument from
part to whole.'

'I have found a mood here,' said Dyver, 'which
seems right, and yet it isn't among that "Barbara,
Celarent" lot; it is A A I in fig. i.'

'You are quite right—it is valid; it is called "a
subaltern mood." There are five of them; they are
really contained in the others. Thus A A I is contained in A A A; for if I can get the conclusion,
"all C is A," it contains the conclusion, "some
C is A." If from the premisses "all cats are animals,
all animals can feel," I draw the conclusion "all cats
can feel," much more can I draw the "weakened
conclusion," "some cats can feel."'

That night I had a strange dream. I thought
Mr. Practical told us he had four rooms which he
wished to stock with certain specimens. The names
of the rooms were 'Fig. i.,' 'Fig. ii.,' 'Fig. iii.,' and
'Fig. iv' He gave Dyver a kind of hand fire-screen,
with 'Distribuas medium,' &c. printed in large letters
upon it. He gave me a stout staff, with Truth
engraved on the handle, bidding me to knock on the
head anything that Dyver could not repel without

violence. Then, with minute instructions as to the kind of specimens we were to secure ('moods' was their name), and as to the manner of capturing the right ones and getting rid of the wrong ones (for they were very like one another), and with no small delight on my part at the idea of a liberal use of my staff, and, lastly, accompanied by a keen-scented pointer named 'Asebinop,' we started before daylight, and made for the land of moods.

Upon arriving there we saw hundreds of animals of a most wonderful description—each looked like three gigantic wasps joined together one above another; and there was a perfect hurricane of 'ergos' and 'therefores,' which they kept hissing with human voices. Suddenly 'Asebinop' pointed; Dyver grew pale, and I clutched my staff. Enveloped in a cloud of darkness, a huge Form loomed before us, and we heard a voice as of an irascible major in hot argument saying over and over again, 'All officers are exposed to temptations, no officers are saints, therefore no saints are exposed to temptations.' A low growl from Asebinop, and the mood (for such it was) caught sight of him for the first time, and shuddered visibly. Meanwhile Dyver was holding up his screen persistently, and shouting out ' Dīs-trĭbŭ|-ās mĕdĭ|-um, nēc|,' &c., with painful stress on the scanning, without any effect, however, upon the monster, until he arrivel at the line 'Et non| dīstrĭbŭ|āt nisi| cūm præ|,' &c., when the form uttered

MOOD-HUNTING.

a cry of pain, and seemed unable to move, being fascinated by the screen. 'Now then, Douglas,' cried Dyver, and with a blow of my club I finished the illicit old wretch. After that we despatched an 'illicit minor' in the same way, and after that an 'undistributed middle.' This last was a horrid sight. It was rolling about in an agony of pain, and gasping out, 'I must be right—"all lotions are medicine," and "all potions are medicine;" so lotions and potions are the same; and in swallowing my lotion to cure this indigestion, I must have been right, but the pain is intense!' We heard afterwards that this mood had argued itself into swallowing poison to cure its indigestion. We owed much to ASEBINOP, though sometimes he took no notice of a mood; but Dyver's distinct reading of the rules was a sure test always. We secured many valid specimens, and brought them home, and Mr. Practical read 'Barbara, Celarent,' &c. over them like a roll-call; and when each mood had answered to its name, he praised us highly for having found them all; and, bidding them retire to their several figures, wished them all 'good night,' and retired.

CHAPTER XIX.

SYLLOGISM—CANON OF FIRST FIGURE—REDUCTION.

' I can't understand how you call a syllogism a universal form of thought, when nobody ever speaks in syllogisms,' said I.

'In ordinary writing or speaking,' replied Mr. Practical, 'part of the full form is suppressed, and the order changed; but to examine arguments it is always better to restore them to their full and original form. These shortened forms are called "Enthymemes" (from ἐν and θυμός, because part is suppressed, understood, or kept "in mind," but not expressed). You get three kinds of enthymeme by suppressing either the major premiss, or the minor, or the conclusion. You may say either, "Socrates is a man," ∴ "Socrates is mortal," or "All men are mortal, ∴ Socrates is mortal," or "All men are mortal, and Socrates is a man." The order of the first two in conversation would be "Socrates is mortal, because he is a man," and "Socrates is mortal, because all men are so." As to the last, its order would not be changed, and if you wish to see how quickly men supply the missing conclusion (which is "in the minds," ἐν θυμοῖς) go to the most illiterate costermonger and say, "All costermongers are scoundrels, and you are a costermonger," and observe the result. We now pass to

'(C). *The canon of the first figure.* A glance at all the moods in the first figure shows us that they all agree in having a universal major premiss and an affirmative minor premiss. Hence what is called the canon or law of the first figure,

(1) The major premiss must be universal.

(2) The minor   ,,   ,, affirmative.

'There are also "canons" of the other figures,[1] but they are of less importance, and I refer you to other books for them. The use of this first figure canon is that by it we are enabled to be still more sure of the validity of the moods in the remaining figures. By a change that does not interfere with their validity we can *reduce* moods of figs. ii., iii., and iv. to fig. i., when we can reassure ourselves as to their validity by applying to them the canon of the first figure. Take CESARE (*i.e.* mood E A E in fig. ii.).

E.   No A is B   We want to bring this mood in
A.   All C is B   fig. ii. to a mood in fig. i., *i.e.*
E. ∴ No C is A   we want to make the middle term come as subject, predicate, instead of predicate, predicate (for herein consists the difference of figure). Remember conversion is our great instrument here. If in the major premiss "No A is B" we know by simple conversion that "No B is A," and the reduction is complete, we have

·[1] Appendix A (iii.).

K

¹ E.   No B is A ⎤ So the mood E A E in the se-
                    ⎥ cond figure becomes the mood
A.   All C is B ⎥ E A E in the first figure, or Ce-
E. ∴ No C is A   ⎦ sare becomes Celarent.

Now apply our canon and we find it obeyed, and we
thus have an additional proof of the validity of
Cesare. Take FESAPO (*i.e.* E A O in fig. iv.) :—

¹ E.   No A is B   ⎤ We want the middle term
                     ⎥ (B), subject, predicate, in-
A.   All B is C   ⎥ stead of predicate, sub-
O. ∴ Some C is not A ⎦ ject. We must change

both premises. By simple conversion we make
"No B is A," but we must not convert an A proposition simply, we must employ conversion per accidens, and it becomes "Some C is B." We now have—

E.   No B is A   ⎤ Here the mood has changed
                     ⎥ as well as the figure, and
I.   Some C is B   ⎥ FESAPO (or E A O of the
O. ∴ Some C is not A ⎦ fourth) becomes FERIO

(or E I O of the first). Apply the canon and you
will find it is obeyed.

'Sometimes you have to change the premises
when you do so *remember you change your major and
minor terms.* Take A E E of fig. ii. (Camestres).

A. All A is B	No C is B	No B is C ......E
E. No C is B	All A is B { but E converts simply, hence :— }	All A is B ......A
E. ∴ No C is A	∴ No C is A	∴ No A is C... E
		∴ No C is A (by simple conversion.)

Apply your canon and you will find it obeyed.

¹ See note at the end of the chapter.

'Thus Camestres becomes Celarent. As you must have noticed, the first letter in each of these strange names in figs. ii., iii., iv. means that the mood when reduced becomes a mood which begins with the same letter in fig. i. The vowels of course mean the mood. The letter " s " after the vowel means " simple conversion," and " p " per accidens in the process of reduction. " M " (mutation) means change the premisses, and " k " means " per impossibile." All these facts might have been gathered from observation and experiment, but it is a help to memory to learn them thus as well.'

'What does "per impossibile" mean? asked Dyver.'

'Reduction, as above exhibited, is called ostensive or simple reduction, but there are two moods, BAROKO (second figure) and BOKARDO (third figure), which are beyond the power of simple reduction. The process by which they are proved to furnish us with true conclusions is as follows (and here you had better not attempt ciphers, but remember these instances and that will be quite enough). If the conclusion they give is not true, it must be false. Assume it to be false, and work on the assumption till you are brought face to face with an obvious absurdity, and then retrace your steps saying, " After all our original conclusion was *not* false, but true." (i.) Take BAROKO (A O O, fig. ii.).

A.   All bantams are fowls, ⎫ If I declared this
O.   Some birds are not fowls, ⎪ conclusion to be un-
O. ∴ Some birds are not ⎪ true, an opponent
        bantams.   ⎭ might reply, " Very
well, assume it false, and you will find you get a new
conclusion, which contradicts one of your premisses,
which, of course, you start by assuming to be true."
If " Some birds are not bantams " is false, its contradictory, " All birds are bantams " is true. Substitute
this new truth for your original minor premiss, and
you get

All bantams are fowls,

All birds are bantams,

(from which two premisses it follows that)

∴ All birds are fowls.

' But this new conclusion contradicts our old
minor premiss " Some birds are not fowls," which
we assumed to be true. Therefore we have come to
an absurdity, as we must do in any case if we
assume our first conclusion, " Some birds are not
bantams," to be false.

' Therefore that first conclusion is true.

' (ii). So with *BOKARDO* (O A O, fig. iii.). Only
here you substitute your new truth for the original
major premiss :—

    O. Some policemen are not fools,

    A. All policemen are in the Queen's service,

    O. ∴ Some of those who are in the Queen's ser-
        vice are not fools.

'The new truth here will be, "All in the Q. S. are fools." Substitute this for the original major, and we get—

> All in the Q. S. are fools,
> All policemen are in the Q. S.,
> ∴ All policemen are fools ;

and this new conclusion is antagonistic to our old major premiss, "Some policemen are not fools," a contradiction. Therefore our original conclusion was not false but true. For by the law of excluded middle a thing must be either true or false, and if this conclusion isn't false it must be true.'

Note.—Putting words for the ciphers above we should reduce thus :

E. No colliers can manage balloons,
A. All good aeronauts can manage balloons,
E. ∴ No good aeronauts are colliers.

⎫
⎬ or fig. ii. may be reduced to fig. i. thus :—
⎭

E. No people who can manage balloons are colliers,
A. All good aeronauts can manage balloons,
E. ∴ No good aeronauts are colliers.

E. No flirts are true women,
A. All true women are dear to men,
O. ∴ Some of the things dear to men are not flirts.

⎫
⎬ or fig. iv. may be reduced to fig. i. thus :—
⎭

E. No true women are flirts,
I. Some (*i.e.* one of the) things dear to men are true women,
O. ∴ Some of the things dear to men are not flirts.

CHAPTER XX.

TRAINS OF REASONING—SORITES.

'SYLLOGISMS may be "heaped" one above another (σῶρος, a heap) in a train of reasoning, called Sorites.[1] Thus :—

>All A is B,
>All B is C,
>All C is D,
>All D is E,
>∴ All A is E.

'This you may resolve into as many syllogisms as there are premisses between the first premiss and the conclusion. Always start with the second premiss and the rest is easy—

All B is C,	All C is D,	All D is E,
All A is B,	All A is C,	All A is D,
∴ All A is C.	∴ All A is D.	∴ All A is E.

'These syllogisms are called pro-syllogisms or epi-syllogisms, according as you regard them as prior or subsequent to one another. There are two rules :—

(1) Only one premiss (the first) can be particular.
(2) Only one premiss (the last) can be negative.

[1] Appendix B.

'For if (1) be violated, you will find when you expand your sorites into its full form that you have a particular major in the first figure (contrary to canon), and if (2) be violated you will find in the same way that you have a negative minor in the first figure (contrary to canon).

' [If asked (as possibly you might be) what is the regressive or Goclenian Sorites, remember it is the reverse of the above. Begin with the last premiss and write the train from last to first, and keep the old conclusion, *e.g.*,

All D is E,

All C is D,

All B is C,

All A is B,

∴. All A is E.

'The rules are reversed, too; only one premiss particular, the last; only one negative, the first (the same propositions being negative and particular as before, you observe)].'

'Would it be enough,' asked I, 'if you were asked about the Goclenian Sorites to give a full description of the ordinary Sorites, and then finish up by saying, "Such is Sorites; and the Goclenian is *not* this, but the reverse," leaving the examiners to draw upon their imaginations for your meaning.'

' It would be a great deal better than leaving the question out altogether,' he replied with a laugh.

CHAPTER XXI.

HYPOTHETICAL SYLLOGISMS.

'As yet our propositions have been only simple or categorical. There are propositions which link together a couple of simple propositions as simple propositions link together a couple of terms. These are called " complex or hypothetical propositions." " Hypothetical" means " with something put under or *supposed* " (ὑπὸ τίθημι = sub-pono = suppose), or " with a condition." Your father says, " I'll give you a horse," that's one thing; but it is quite another thing if he says, " I'll give you a horse IF you pass your examination." The universal dislike of IF's is proverbial, for they make all the difference being the signs of "conditions." Now simple propositions may be linked together in two ways : first, where the truth of the consequent depends upon the truth of the antecedent (antecedent and consequent are the names of the two simple propositions when linked together; the first, the antecedent; the second, the consequent) ; and secondly, where the truth of the consequent depends upon the falsity of the antecedent. The first kind are called " conjunctive," *e.g.*, "If the weather is rainy my sponge is damp " (their sign is "if "). The second kind are called disjunctive, *e.g.*, " Either Logic is deep or I am dull " (their sign is " either—or "). Hence we may divide propositions in the following manner :—

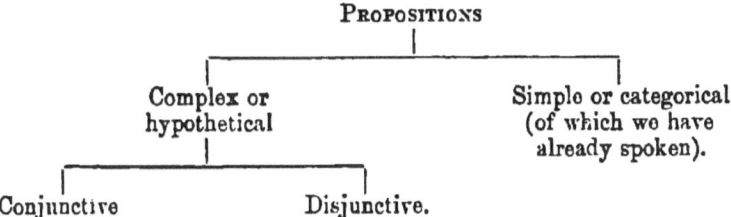

PROPOSITIONS

Complex or hypothetical

Simple or categorical (of which we have already spoken).

Conjunctive   Disjunctive.

' Now syllogisms are composed of these hypothetical propositions, and so borrow their names. We have conjunctive and disjunctive hypothetical syllogisms, *i.e.* syllogisms composed of such premisses. Cases where both premisses are hypothetical we shall discuss under " dilemma." For the present we shall consider cases where one premiss is hypothetical and one simple.

' I. *Conjunctive Hypothetical Syllogisms.* These admit of two valid conclusions out of the four possible ones you get by affirming and denying the antecedent and consequent.

(*a*) If the weather is rainy, my sponge is damp.

The weather *is* rainy,

∴ my sponge is damp.

Affirming the antecedent for a minor premiss.

(*b*) If the weather is rainy, my sponge is damp.

The weather is *not* rainy.

No conclusion.

Denying the antecedent for a minor premiss.'

'Surely,' said I, 'it follows that "my sponge is not damp."'

'No,' said he, 'for it may have fallen into the bath. Suppose I say "If the 'Brighteyes' are going, I shall enjoy the ball;" it does not follow that if they do not go, I shall not enjoy the ball, for I may meet the "Lighttoes."'

'(c) If the weather is rainy, my sponge is damp.

My sponge is damp.

No conclusion.

Affirming the consequent for a minor premiss.

For though it follows that my sponge is damp if the weather is wet, it does not follow that the weather is wet because my sponge is damp. If the wife weeps because the husband is condemned to death, does it follow that the husband is condemned to death *because* the wife weeps?

'(d) If the weather is rainy, my sponge is damp.

My sponge is not damp.

∴ the weather is not rainy.

Denying the consequent for the minor premiss.

Thus the conjunctive hypothetical syllogism admits of two conclusions—where you affirm the antecedent called "constructive," and where you deny the consequent called "destructive."

II. *Disjunctive Hypothetical Syllogisms* admit of four conclusions, for you may affirm or deny antecedent and consequent.

e.g., (1) Either Logic is deep, or I am dull.
       Logic is deep.
       ∴ I am not dull.

(2) Either Logic is deep, or I am dull.
       Logic is not deep.
       ∴ I am dull.

(3) Either Logic is deep, or I am dull.
       I am dull.
       ∴ Logic is not deep.

(4) Either Logic is deep, or I am dull.
       I am not dull.
       ∴ Logic is deep.

'The *dilemma* is a combination of conjunctive and disjunctive premisses. As it is a difficult matter to understand, I should advise you to remember the three forms of it by their examples, and after your examination go more deeply into the theory of it as expounded in the books on Logic. There are :—

(i.) The simple ⎫ constructive ⎧ Instance: "Science."
(ii.) The complex ⎭ dilemma ⎩ Instance: "Politician."
iii.) The destructive dilemma.   Instance: " Jesting at Scripture."

' (i) The *Simple Constructive Dilemma* is of this form :—

If science lightens labour, it should be culti-
    vated; and if science invigorates the faculties,
    it should be cultivated;
But science does one, or the other;
Therefore science should be cultivated.

' (ii) The *Complex Constructive Dilemma* is of this form :—

> If a politician (who finds he is wrong) changes his views, he is inconsistent; and if he does not change them, he is not conscientious;
>
> He must either change them or not change them;
>
> Therefore he must be either inconsistent or not conscientious.

' (iii) The *Destructive Dilemma* is of this form :—

> If a man were wise, he would not jest at Scripture in fun; and if he were good, he would not do so in earnest;
>
> He must do it either in fun, or in earnest;
>
> Therefore he must be either not wise or not good.

These three (the substance of which is borrowed from Mr. Jevons's book) will be quite enough to show that you understand what dilemma means. A dilemma may be rebutted thus : 1st Dilemma. "Do not enter into public affairs; for if you say what is just, men will hate you; and if you say what is unjust, the gods will hate you. You must do one or the other; therefore you must be hated by gods or by men." 2nd Dilemma. "Do enter into public affairs; for if you say what is unjust, men will love you; and if you say what is just, the gods will love you; therefore you must be loved by gods or men." '

' In the "Oxford Spectator" I saw a dilemma,' said I, ' which seemed conclusive. How would you

rebut this : " Examinations are useless; for if you
know the questions already, they teach you nothing;
and if you do not know the questions, they teach you
nothing ; you must either know them or not know
them ; therefore examinations are useless ? " '

To which Dyver replied, to my astonishment:
' Examinations are useful ; for if we do know the
questions, they teach us how to express our knowledge ; and if we do not know the questions, they
teach us what our weak points are (and it is not
their fault if we do not remedy them) ; we must
either know the questions or not know them ; therefore examinations must be useful.'

Mr. Practical expressed his approval, and I
began to think that if Mr. Practical taught Dyver
much more it would be a case of the young horse
running away with the ' coach ' altogether.

CHAPTER XXII.

PROBABLE REASONING.

' HITHERTO we have spoken of syllogisms with strictly logical forms of propositions for their premises, when the conclusions are certain.　There are two kinds of reasoning which furnish us with conclusions not strictly certain, but of sufficient weight to influence our actions in life—self-infirmative, and self-confirmative inference.　By help of such probable reasoning we are enabled to make syllogisms out of propositions with the sign "most" or "many," instead of "all" or "some;"　and we can take into account the force of such words as "probably," without (as in strict Logic) thrusting them into the subject or prædicate.

' (i.) *Self-infirmative inference* is where each fresh fact weakens the conclusion, so that the more premisses you have the less likely is it that the conclusion will be true.　"Never go out of doors in a severe frost," says the anxious mother, " because the Humane Society's men are obliged to drink."

' But what has that to do with my "going out?" you ask.

' " You know," she replies, " some of those who go out in such times are tempted to skate, and some of those who skate break the ice, and some of those who break the ice are rescued by the Humane Society's men, and some of these men drink to keep themselves warm ; therefore, some of the men who venture out in a severe frost may have to be rescued by possibly intoxicated men."

'You laugh at this argument, because every fresh premiss weakens the conclusion.   So with the words " possibly," " probably," &c.

' By this process you can argue that men with money are likely to commit suicide ; thus :—

Men with money probably invest it ;

Men who invest probably speculate ;

Men who speculate possibly lose all ;

Men who lose all are probably pinched with poverty ;

Men who are pinched with poverty probably despair ;

Men who despair possibly commit suicide.

∴ To this extent men with money are likely to commit suicide.

The amount of the probability may be estimated to a fraction ; but calculations of this sort seem to belong rather to mathematics than Logic.'

' What a comfort ! ' I reflected.

' (ii.) *Self-confirmative inference* is where each fresh fact strengthens the conclusion you wish to

establish. It is called "circumstantial evidence,"
or a "chain" or "coil" of evidence. Many verdicts
are awarded in the Law Courts solely upon the
strength of this evidence, which amounts in some
cases almost to certainty. Each fact may be regarded
as the minor premiss of a syllogism, with a probable
major premiss and a probable conclusion. Given the
assertion "That our cook gave the joint to a follower.
when she *said* the dog ate it." It is required to
prove this assertion by circumstantial evidence—for
nobody actually saw her give it. The evidence is as
follows:—"The dog was a remarkably well-behaved
dog." "On the day of the mysterious disappearance
the kitchen blinds were kept down." "Certain other
articles (such as beer, tea, spirits, &c.), which dogs
would not eat or drink, vanished during the same
cookship," &c. Each of these facts becomes the
premiss of a syllogism; thus:—

I.

Well-behaved dogs probably are innocent of a
    theft.

This was a well-behaved dog.

    ∴ The dog is probably innocent of the theft.

II.

The drawing of the blinds probably betokened
    the presence of a follower.

The blinds were drawn.

    ∴ A follower was probably present.

If a follower was present, probably the dog is
innocent of the theft.

A follower was probably present.

Therefore the dog is probably innocent of the
theft.

III.

If tea, spirits, &c. disappeared also, the dog
is probably innocent of the theft.

Tea, spirits, &c. did disappear.

Therefore the dog is probably innocent of the
theft.

Every syllogism has the same conclusion, you
observe; and every fact thus becomes a link in the
chain of evidence. You may find instances for yourselves in the newspapers. This evidence is also
called " Self-corroborative," from "robur," strength.

'And here I may mention an exception to the
rule—" Ex duobus particularibus nihil sequitur," or
"two particular premisses prove nothing" (Rule 4
of the Syllogistic Rules); for there is one case in the
third figure where two particular premisses do prove
something; e.g.,

" Most pins are good ;
Most pins are cheap.
∴ Some cheap things are good."'

CHAPTER XXIII.

THE FALLACIES.

'A FALLACY (fallo, to deceive) is an argument which seems to be true, but is really false. The fallacies have been classed under two heads: material (extra dictione) and formal (in dictione), *i.e.* fallacies arising from mistakes in the matter and the form. For the mistakes in matter Logic is not responsible (see "Matter and Form of Thought"). But we shall adopt Mr. Fowler's arrangement.—(N.B. Learn off this scheme, and read through the explanation that follows).

I. *Assumption of a false premiss.*
   Achilles and the tortoise.

II. *Neglect of the laws of deductive inference.*

   Illicit major.[1]
   　,,　minor.
   Undistributed middle.
   Petitio principii (begging the question).

III. *Ignoratio elenchi, or irrelevancy.*

   Argumentum ad hominem...(" early rising ").
   　　,,　　,, populum...(" weeping wife").
   　　,,　　,, baculum...("religious persecution ")

---

[1] For several instances of these see Appendix A (ii).

IV. *Ambiguity of language.*

Ambiguous (analogous, equivocal, &c.) words...(" box, muzzle, &c.").

Composition and division...(" 3 and 2—odd and even ").

Fallaciæ accidentis [2]...(" mad dog's bite ").

Paronymous terms...(" drunk once, drunk ever ").

Fallacia plurium interrogationum...(" Have you left off poaching yet ? ").

Amphibolia, or amphibology...(" The duke yet lives that Henry shall depose ").

Fallacy of accent...(" and they saddled *him* ").

' (I.) Is a mathematical difficulty, and strictly speaking is out of the sphere of Logic. A tortoise starts so many yards ahead of the fleet Achilles, and, according to mathematical calculations, Achilles never overtakes the tortoise. For when the tortoise has advanced a few yards, Achilles has gained so much that there is only half the distance between them, and presently only a quarter, and then one-eighth, and then one-sixteenth, and so on, the fraction becoming smaller and smaller, but never quite vanishing, so Achilles could never (according to this) catch the tortoise quite. The logicians in despair cried, " Solvitur ambulando," or " Walk it and see ; " a better answer would have been, " Logic has nothing to do with the complications and difficulties and errors of the matter." (See " Form and Matter of Thought ").[1] Under this head come all fallacies of the matter.

' (II.) These have already been explained except the petitio principii. " Begging the question " is proving the conclusion you wish to establish by

[1] P. 26 and p. 29.

assuming that conclusion to be true before you have proved it to be true. There is the simple form of it. "A man is a man," says a lady. "Why? prove it," say you. "Because he is a man," she retorts. It may be by a synonym (a like name). "The air is dry." "Why?" "Because the atmosphere is dry." Under this head comes the "argument in a circle," for you move in circles when you prove your statements by making them your premisses as well as your conclusion, e.g., "Fire is hot, because it burns," says your lecturer, and presently, when asked, "But why does fire burn?" he blandly replies, "Because it's hot, of course." In long speeches such arguments pass unnoticed, especially in a language like the English, compounded of Saxon, Norman, and Latin—where there are so many different words all meaning the same—forming excellent covert for fallacies of this kind to hide themselves in.

'III. Ignoratio elenchi, or irrelevancy, means ignorance of the exact point to be refuted. Your arguments are irrelevant when they miss the point to be assailed, and batter down some unoffending, innocent, unguarded point. Argumentum ad hominem misses the theory and assaults the man. Its object is to find a flaw, not in the theory advocated, but in the person advocating. When Mr. Gladstone advocated the theory of Disestablishment of the Irish Church opponents condemned the idea simply because its holder "once thought differently," in other

words, galloping by the fortified heights of the
theory, they levelled their batteries at the weak
flesh from whence it emanated. They made the
inconsistency of the person a proof of the inefficiency
of the measure. And so when a late riser speaks in
glowing terms of early rising and its benefits, people
laugh. The point to be assailed may be weak or
strong, but the Logician must not swerve from that
point.'

'But when a holder is attacked instead of his
theory,' said Dyver, 'do we not often find that a
man says one thing and does another, and ought we
not rather to believe what he does than what he
says?'

'True; but we cannot "practice what we preach."
As a Logician you have only to ask, "Is the theory
good or bad?" and all questions of persons should
be set aside. The argument "ad populum" appeals
to the feelings rather than the judgment of an
audience. When a prisoner at the bar brings his
weeping wife and children to excite pity he is missing
the point, so to speak; for the point he ought to aim
at is to prove his own innocence, the point he does
aim at is to win their pity. The argument "ad
baculum," declining to face the argument, lays hold
upon the voice that pleads it, and by force compels
it to hold its peace. It is the refutation of those who
are strong of hand, but weak of head, for it crushes
theories by brute force. Religious persecution is an
instance.'

'That reminds me,' said I, 'of the fox and the serpent and the man. There was once a man who found a wounded serpent and put it into his bosom. And as soon as the serpent revived it bit the man's bosom. And the man called it an "ungrateful monster;" upon which a long argument arose between the two as to whether the man or the serpent was the more ungrateful animal on the whole. Appeals were made to various animals (old cows, hunters, &c.) without any abatement of the hot discussion, until the fox was consulted. "If you wish this fierce argument to come to an end," said the fox to the serpent, "be so good as to enter into this bag, and you will be so convinced that you will never utter another syllable." Upon the serpent's ready obedience, the fox tied up the neck of the bag, and handing it to the man said, "Stamp on him," and so the argument was ended.' Mr. Practical was highly pleased at this, and continued.

'IV. All fallacies arising from ambiguous use of language. Equivocal and analogous words are instances. Remember this—*any word may be used in one sense or more than one sense.* If it is used in one sense, it is said to be *univocally* used, as "butter" (there are very few of these). If in more than one sense, there may be a connection between the several senses, or there may not. If there is, it is said to be *analogously* used, as "muzzle," in its three senses of "part of a gun," "part of a dog," "a gag" (all have something to do with the mouth). If there is not, it is said to be *equivocally* used, as box in its three senses of a "chest," a "tree," a "blow" (no connection between these). Hence ambiguous middles in syllogisms, as "The people here are the church; the stones built up here are the church; .˙. the people here are the stones built up here."

'The fallacy of composition is arguing from what is true of things taken separately to what is true of things taken together; *e.g.*, from "two and three are even and odd" to argue "five is odd and even." And division is the converse, from "five is odd" to argue "two and three are both odd." Or from "Jones and his wife were happy when single" to argue "Jones and his wife are happy together; or from "Jones and his wife are miserable together" to argue that "Jones and his wife were miserable when single."

'The fallaciæ accidentis are two, (1) the argument "a dicto simpliciter ad dictum secundum quid;" (2) the argument "a dicto secundum quid ad dictum simpliciter." Which formidable expressions simply mean arguing from what is true as a general rule to what is true under particular circumstance, and from what is true under particular circumstances to what is true as a general rule.

'Now as a general rule "it is a bad thing to cut off a man's arm," but under certain circumstances (bite of a mad dog, for instance) it is not bad. According to the first of the above fallacies, one would persist in saying, " Do not cut off his arm—never mind what has happened—do not cut it off, for it's a bad thing for a man to have his arm cut off." According to the second fallacy one would say, " Let us cut off all the arms of all men, for once upon a time I knew a man (who had been bitten by a mad dog) who gained great advantage from losing his arm."

'The fallacy of paronymous terms is exemplified when you say, " Job Turnips is a drunkard," because he has been once or twice drunk ; or "a thief," because he has once or twice stolen.

' The fallacia plurium interrogationum is the fallacy of asking two questions in one. For instance, take the question of a barrister examining a prisoner : " Have you left off poaching yet ? " when the unfortunate man criminates himself by saying, " Lor' bles yer, yes, this six months or more." For the barrister asked him two questions in one : " Did you ever poach ? " and " Have you left off yet ? " meaning only to discover whether the man ever poached at all.

' The fallacy of amphibolia, or amphibology is a purely grammatical one, and is the name given to such doubtful passages as " The duke yet lives that Henry shall depose." Where you do not know whether it is the duke who shall depose Henry, or Henry the duke.

'The fallacy of accent is a mere matter of emphasis. For instance, if you lay stress upon the pronoun "him" in the following words, " Saddle me the ass. And they saddled *him*," you convey a wrong impression.

Lastly, there are two fallacies not mentioned in the scheme above; that of " non-causa' procausâ" (as to say, " A comet shone, therefore there will be war "), where you take for a cause that which is not really so;[1] and the fallacy of the consequent (fallacia consequentis or " non-sequitur "), *e.g.*, " he is an animal ; *ergo* he is a man."

[1] This fallacy is the one referred to by the words, 'Post hoc, sed non propter hoc,' as the man said when he stumbled after partaking freely of a well-known light wine.

APPENDIX A.

(i.) WITH regard to the four figures in the Syllogism remember that—

Fig. I. is most useful for the discovery or proof of properties of a thing.

Fig. II. is most useful for the discovery or proof of distinctions between things.

Fig. III. is most useful for the discovery or proof of instances or exceptions.

Fig. IV. is most useful for the discovery or exclusion of the different species of a genus. (Lambert.)

And this is because :—

(1) Fig. I. can prove (*i.e.* have for its conclusion) A, E, I, or O.

(2) Fig. II. can prove (*i.e.* have for its conclusion) E and O only (*i.e.* negatives only , for as in this figure the middle term is prædicate (*i.e.* prædicate in both premisses) unless one of the premisses were negative, the middle term would be undistributed.

∴ One of the premisses must be negative.

∴ The conclusion must be negative.

(3) Fig. III. can prove (*i.e.* have for its conclusion) I or O only (*i.e.* particulars only).[1]

---

[1] By working out the six moods in Fig. III. you will find an illicit minor in each case if you make the conclusion universal. For practice work them out. See Appendix A (iv.).

(4) Fig. IV. proves E, I, O, and is of little use because its work can be done more easily by Fig. I.

N.B.—From this it is plain that Fig. III. is the best for the purpose of overthrowing an adversary's conclusion, as it furnishes us with material for the establishment of a contradictory opposition (see p. 104).

(ii.) There are three rules (4), (6), and (8), which require further explanations :—

Rule (4). 'Two particular premisses prove nothing.' I and O are the only forms of particular propositions. Two particular premisses must therefore be either I, I; or O, O; or I, O; or O, I.

Take these four combinations in all figures :—

Fig. i.	Fig. ii.	Fig. iii.	Fig. iv.
I. Some B is A	Some A is B	Some B is A	Some A is B
I. Some C is B	Some C is B	Some B is C	Some B is C
No conclusion.	No conclusion.	No conclusion.	No conclusion.

For by [1] ASEBINOP it is plain that the Middle Term (B) is not once distributed in any of these—∴ There can be no conclusion.

Fig. i.	Fig. ii.	Fig. iii.	Fig. iv.
O. Some B is not A	Some A is not B	Some B is not A	Some A is not B
O. Some C is not B	Some C is not B	Some B is not C	Some B is not C
No conclusion.	No conclusion.	No conclusion.	No conclusion.

In all these there are two negative premisses, which give no conclusion (p. 130).

Fig. i.	Fig. ii.	Fig. iii.	Fig. iv.
I. Some B is A	Some A is B	Some B is A	Some A is B
O. Some C is not B	Some C is not B	Some B is not C	Some B is not C
No conclusion.	No conclusion.	No conclusion.	No conclusion.

In all these (by Rule (5) p. 130) the conclusion, if there be any, must be *negative*, and by Rule (6) *particular*.

In all four figures it will be '∴ *Some C is not A.*'

---

[1] For *Asebinop*, see pp. 78 and 126. It only means that an A prop. distributes its subject, E both subject and predicate, I neither, O predicate.

By ASEBINOP we know all four to be cases of Illicit Major (p. 124–5), *i.e.* the term A is used in its full sense in the conclusion, but only in a partial sense in the premisses – and this is arguing from part to whole.

O. Some B is not A	Some A is not B	Some B is not A	Some A is not B
I. Some C is B	Some C is B	Some B is C	Some B is C
No conclusion.	No conclusion.	No conclusion.	No conclusion.

For here we find that in figs. i. and iii. the Middle Term (B) is undistributed ; and in figs. ii. and iv. there will be an ' Illicit Major ' as above.

N.B.—' No conclusion ' means ' No valid conclusion.'

Rule (6). ' If either Premiss be particular, the conclusion must be particular.'

For, so to speak, universal conclusions are expensive luxuries to their premisses ; premisses of small quantity cannot afford them. To keep up a universal conclusion both premisses must be at their full or universal stage in point of quantity, and any diminution at once tells sensibly upon their conclusion. Two universal premisses can sustain a universal conclusion ; but impoverish one of your premisses, without making a corresponding reduction in your conclusion, and you will find that Illicit Major or Minor has eaten away, like a canker, the truth of your Reasonings. This is best seen by going through all possible combinations of Particular and Universal Premisses (the combination of two particulars is by Rule (4) invalid):—

These are $\left(\begin{smallmatrix} A \\ I \end{smallmatrix}\right)$, $\left(\begin{smallmatrix} A \\ O \end{smallmatrix}\right)$, $\left(\begin{smallmatrix} I \\ A \end{smallmatrix}\right)$, $\left(\begin{smallmatrix} O \\ A \end{smallmatrix}\right)$, $\left(\begin{smallmatrix} E \\ I \end{smallmatrix}\right)$, $\left(\begin{smallmatrix} E \\ O \end{smallmatrix}\right)$, $\left(\begin{smallmatrix} I \\ E \end{smallmatrix}\right)$, $\left(\begin{smallmatrix} O \\ E \end{smallmatrix}\right)$. Of these $\left(\begin{smallmatrix} E \\ O \end{smallmatrix}\right)$ and $\left(\begin{smallmatrix} O \\ E \end{smallmatrix}\right)$ are invalid by Rule (3), p. 120. We have now to show that none of the remaining six moods, or parts of moods, can have a universal conclusion (*i.e.* A or E) in any of the figures.

Fig. i.	Fig. ii.	Fig. iii.	Fig. iv.
A. All B is A	All A is B	All B is A	All A is B
I. Some C is B	Some C is B	Some B is C	Some B is C

Here in figs. ii. and iv. the middle is undistributed (ASEBINOP), and there is no conclusion. In figs. i. and iii. the conclusion cannot be E by Rule (8) (which will be explained below); and it cannot be A, or in both figures you would have Illicit Minor.

Fig. i.	Fig. ii.	Fig. iii.	Fig. iv.
A. All B is A	All A is B	All B is A	All A is B
O. Some C is not B	Some C is not B	Some B is not C	Some B is not C

In all these the conclusion, if universal at all, must be universal neg. (E) by Rule (5). Now supposing E to be the conclusion in each, we have in fig. i. Illicit Minor and Major together (ASEBINOP); in fig. ii. Illicit Minor; in fig. iii. Illicit Major; and in fig. iv., the middle being undistributed, there can be no conclusion at all.

I. Some B is A	Some A is B	Some B is A	Some A is B
A. All C is B	All C is B	All B is C	All B is C

In all of these the conclusion, if universal at all, must be A by Rule (8). In figs. i. and ii. there are no conclusions at all, the middle being undistributed. In figs. iii. and iv. if the conclusion be A, you get illicit minors in both.

O. Some B is not A	Some A is not B	Some B is not A	Some A is not B
A. All C is B	All C is B	All B is C	All B is C

Here in fig. i. there is no conclusion (undist. mid.). In the rest, the conclusion, if universal at all, must be E (Rule 5), which would involve, in fig. ii. Illicit Major; in fig. iii. Illicit Minor; in fig. iv. both together (ASEBINOP).

E. No B is A	No A is B	No B is A	No A is B
I. Some C is B	Some C is B	Some B is C	Some B is C

In all these the conclusion, if universal at all, must be E (Rule 5). This would involve an Illicit Minor in all four figures (ASEBINOP).

| I. Some B is A | Some A is B | Some B is A | Some A is B |
| E. No C is B | No C is B | No B is C | No B is C |

In all these, the conclusion, if universal at all, must be E (Rule 5). This would involve Illicit Major in all four figures (ASEBINOP).

Rule (8). ' Let not the conclusion be negative unless one of the premisses is negative.'

The necessity of this is evident from a consideration of the principle upon which all Syllogism depends. For in it two things are compared with one another through the medium of a third thing. The comparison of the 1st thing with the 3rd thing makes one premiss; the comparison of the 2nd thing with the 3rd thing makes another premiss; and the comparison of the 1st and 2nd thing which results is the conclusion. Now if this conclusion expresses dissimilitude, *i.e.* is negative, it is clear that in one or other of the premisses there must have been dissimilitude expressed before, else whence could this dissimilitude have sprung? for the conclusion is only a summary of what was contained in the premisses; in other words, if the conclusion is negative (*i.e.* expresses dissimilitude between its subject and prædicate) one of the premisses must have been negative before it.

N.B.—The following are instances in *words* as opposed to *ciphers.*

Violations of (Rule 4) :—

'Fishes fly; fishes do not fly. This proves the futility of all knowledge.'

Fig. I.
I. Some fishes fly.
I. Some things that don't fly are fishes.
No conclusion.

'Lancashire is lovely, old women are lovely too. So Lancashire must be an old woman or Logic is untrue!'

Fig. II.
I. Some of Lancashire is lovely.
I. Some old women are lovely.
No conclusion.

'Inspectors are stern, and Inspectors are mild, at the same time and place. So " stern " and " mild " must be the same thing.' = Fig. iii.
I. Some inspectors are stern.
I. Some inspectors are mild.
No conclusion.

'Some ladies must be addicted to drink, for they are thirsty, and thirsty people are often addicted to drink.' = Fig. iv.
I. Some ladies are thirsty.
I. Some thirsty people are addicted to drink.
No conclusion.

'Some eagles have no wings; for all birds are not eagles, and there are other winged things besides birds.' = Fig. i.
O. Some birds are not eagles.
O. Some winged things are not birds.
No conclusion.

'Silence and speech are the same things; for sometimes one is not seemly, sometimes the other.' = Fig. ii.
O. Some silence is not seemly.
O. Some speech is not seemly.
No conclusion.

'Carpenters can't be handsome, for tall men are often handsome, and it isn't every carpenter that is tall.' = Fig. i.
I. Some tall men are handsome.
O. Some carpenters are not tall men.
No conclusion.

'Some cabmen are honest, for all men can't be dishonest, and cabmen are men.' = Fig. iii.
O. Some men are not dishonest.
I. Some men are cabmen.
No conclusion.

Violations of (Rule 6) :—

'We must eat cook's children, for we eat cook's productions; dumplings are her productions, and we eat them.' = Fig. i.
A. All dumplings are to be eaten.
I. Some of cook's productions are dumplings.
∴ All cook's productions must be eaten.

'How wretched must all men under petticoat government be! For all henpecked men are under petticoat government, and how miserable are some henpecked men!' = Fig. iii.
I. Some henpecked men are miserable.
A. All henpecked men are under petticoat government.
∴ All under petticoat government are miserable.

N.B.—In both of these a particular conclusion would have passed as valid.

(iii.) Rules of fig. ii. are
$$\begin{cases} \text{The Major Premiss must be} \\ \text{universal.} \\ \text{One of the Premisses must be} \\ \text{negative.} \end{cases}$$

fig. iii. „
$$\begin{cases} \text{The Minor Premiss must be} \\ \text{affirmative.} \\ \text{The conclusion must be} \\ \text{particular.} \end{cases}$$

Fig. iv. can be more clearly arranged as fig. i.

(iv.) The moods of fig. iii. are Darapti, Disamis, Datisi, Felapton, Bokardo, Ferison (see p. 124).

A. All B is A.	I. Some B is A.	A. All B is A.
A. All B is C.	A. All B is C.	I. Some B is C.
I ∴ Some C is A.	I. ∴ Some C is A.	I. Some C is A.
E. No B is A.	O. Some B is not C.	E. No B is A.
A. All B is C.	A. All B is C.	I. Some B is C.
O. Some C is not A.	O Some C is not A.	O. Some C is not A.

In all of these a universal conclusion involves an ' illicit minor.' So the 3rd fig. only gives particular conclusions. Putting two of the above into words, we get:—

DARAPTI.
All fishes live in the water.
All fishes can swim.
∴ Some of the things that can swim live in the water.
} = { It would not be right to infer that 'all things that can swim live in the water,' for even swimming masters do not live there always.

DISAMIS.
Some persons can preach.
All persons think they can.
∴ Some of the persons that think they can preach, can preach.
} = { It would be wrong to conclude that all the persons that think they can preach, can preach.

M

APPENDIX B.

—•◦•—

An instance of Sorites would be :—

All fleas are animals.
All animals sustain life by assimilating food.
All that lives by assimilating food is liable to hunger
All that is liable to hunger may be half famished.

∴ All fleas may be half famished.

Here A = Fleas.
B = Animals.
C = ' Sustaining life by assimilating food.'
D = ' Liable to hunger.'
E = ' May be half famished.'

And the Sorites may be resolved as in the cipher-form

APPENDIX C.

————◦•◦————

Turning back to p. 103 we find inductive inference put aside. Induction or inductive inference starts from particulars and works towards universals. The process is fully explained under the heading 'What is Science?' (pp. 4 to 13; read carefully) for all science is inductive (p. 184, 'Method'). As deduction has its forms and rules (discovered by Logic), so induction has its forms and rules (discovered by Logic); and we have inductive as well as deductive logic. The mind of man is as carefully controlled in its journey (μέθοδος) from the particular facts up to the universal laws, as it is in its journey from the universal law down to the particular facts. (See illustration, p. 45.)   Let us consider:

1. *Induction—what it is.*
2. *The principles upon which induction ultimately depends, as deduction depends upon its principle, p. 113 (A.).*[1]
3. *The processes required for induction* (observation and experiment).
4. *The 'methods' in Induction corresponding to the syllogistic laws in Deduction.*

[1] This principle of syllogism is founded on the three great laws upon which all deductive reasoning depends (Laws of Thought, p. 160). The principles of inductive inference are not founded upon these three self-evident axioms. Consequently we do not find the same certainty about our conclusions in induction as we have in deduction. For the principles of the syllogism are laws, the violation of which is inconceivable, whereas the principles of induction are laws, whereof the violation is not inconceivable. For the principles of induction are generalisations from experience, and not like those of deduction part of our nature, which we found but did not make.

(1) *Induction* has been called an 'argument from the known to the unknown,' or 'generalisation from experience,' or 'inference from particular facts to universals.' (Pp. 4–13 give the process.) For by induction the old man ascends from the particular objects he lets fall to the universal laws of gravity. Most of the laws which deduction brings down to particular facts have been by induction raised from particular facts beforehand. For induction precedes deduction except in cases where the laws were implanted in us by nature, and even then induction in a sense precedes; for by it we *find* those implanted laws (though we do not *make* them) before we can bring them down to particular facts.

Now, induction may be thus subdivided (starting with arguments from the known to the known, and working towards arguments from the known to the unknown)[1] :—

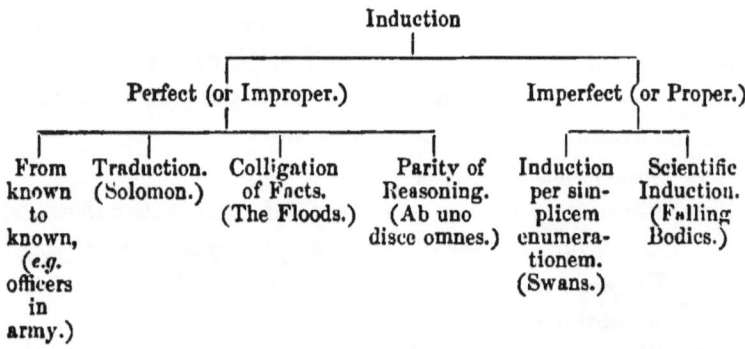

Induction

Perfect (or Improper.)   Imperfect (or Proper.)

| From known to known, (e.g. officers in army.) | Traduction. (Solomon.) | Colligation of Facts. (The Floods.) | Parity of Reasoning. (Ab uno disce omnes.) | Induction per simplicem enumerationem. (Swans.) | Scientific Induction. (Falling Bodies.) |

In arguing from particular facts to universal laws you may know all the particular facts or you may not. If I say all the officers in the British army weigh more than six stone, after having gone through the Army List, and caused each officer to be weighed, I am not in my calculation stating more

---

[1] The four subdivisions of perfect induction are arranged according as they resemble more or less closely imperfect induction. Parity of reasoning might almost be called imperfect induction; colligation is nearer than traduction; traduction and 'from known to known' are the pure cases of perfect induction.

than I had already stated in my particular premisses. But supposing some of our officers were stationed in parts wheie no means of weighing existed, and I persisted in my conclusion, then I should be stating something more in the conclusion than I had already stated in my particular premisses. Then there would be that 'leap in the dark' which, according to some, takes away the certainty of induction and makes it *imperfect* instead of *perfect*; according to others, is the very 'soul' of induction, and makes it *proper*, instead of *improper*. For if you know *all* your particulars, you gain in the certainty of your conclusion while you lose in its originality (so to speak), for it only tells you what you already knew. Whereas if you do not know all the particulars, but under the impulse of what you do know hazard a conclusion for the rest, you gain in the originality of your conclusion, but you lose in its certainty; for the particulars you never saw may be unlike those you did see, and this would upset your conclusion. If you read that each of the kings of England was above four years old when he began to reign, your conclusion to the effect that 'all the kings of England were above four years old when they began to reign,' teems with certainty, but lacks interest. But if, after enumerating your racing experiences, you come to the conclusion that 'all gentlemen of the turf are honourable men,' your conclusion, while teeming with originality (and the bigger the leap the more pleasurable the sensation—for we have as strong a tendency to jump with the mind as crickets and fleas with the body—see pp. 6 and 10), lacks certainty. A merit and a fault, then, attach to each of these two cases; where you do know all the particulars, and where you do not, and so two names have been given to each case, the former being called 'perfect induction' by those who valued its certainty, but 'improper' by those who looked for something more in the conclusion than they already had in the premisses; the latter being called 'imperfect induction' by those who looked for certainty, but 'proper' by those who valued the fresh information given by a conclusion which did something more than re-echo the words of its premisses. For rich discoveries in science we are indebted to induction proper (or imperfect), but we must not forget the less brilliant aid of induction proper (or perfect). The 'leap in the dark'—like riding across unknown country, contrasted with trotting along roads—has its dangers as well as its charms. Glorious discoveries, like that of the steam-engine, are doubtless the results of such a leap in induction proper, but how many a man has bitterly rued the day when, rejecting the safer argument, he chose the more brilliant one, and from the narrow premisses of his happy experience, 'This married man has peace—and that—and that, &c.,' leapt in the dark, cried 'all married men have peace,' and took him a wife !

Perfect (or improper) induction also includes *traduction*, or arguing from particulars to particulars, e.g., Solomon was the wisest man, Solomon was a married man, ∴ the wisest man was a married man. Also *Colligation of facts*, closely resembling the argument from the known to the known as above described, only here our conclusion is a general conception or universal idea rather than a law. Kepler wished to find out the curve described by the planet Mars in its course. He marked all its particular positions, and, when the course was completed, 'read off,' so to speak, an ellipse from his marks. So a man when the floods are out at Oxford, after visiting every side of Oxford is enabled to say 'the floods surround Oxford.' Supposing he did not visit every side of Oxford, and still came to the same conclusion, his induction would be 'proper' instead of 'improper.' Mr. Mill gives the instance of a sailor who views undiscovered land, and after sailing all round concludes that it is an island. Mr. Mill says this is no inference, for it 'brings in' no new truth (p. 103). At all events it 'brings in' truth in a new shape, and, as Dr. Whewell observes, the sailor who joins his disconnected observations in one universal idea, may be as truly said to ' bring in ' something new as the jeweller who strings together the pearls of a necklace; the pearls were all there before, but the uniting thread is new.

Lastly, there is the argument from *parity of reasoning*, e.g., the propositions of Euclid. We take a single instance of a triangle, and show that any two sides of it are together greater than the third, and from this single instance we lay it down as true of all triangles, that any two of their sides are greater than the third. In walking nobody denies that, 'cæteris paribus,'[1] it is a short cut if you can cut off a corner. Now here we argue from the known to the unknown, for no one has seen all triangles, and yet there is such a strong likeness between all triangles, that in a sense we know all if we know one; ' Ab uno disce omnes.' Hence there is so much certainty about this inference that it ranks as a perfect induction. *Example* (p. 180) to be a safe inference, must only deal with cases where a strong resemblance exists between the particular instances, as in geometrical figures.

We now come to *induction proper* (*or imperfect*). This is an argument from the known to the unknown, as well as from the particular to the universal. ' This and that and the other swan—all the swans we've seen or heard of' (Socrates might have said) ' are white. ∴ All swans are white.' ' This and that and the other body, when let fall, fall towards the earth, even though, from the revolutions of the earth, they have to fall upwards. ∴ All bodies fly towards the earth.' ' This and that and the other application of fire to powder result in explosion. ∴ All do.' And now to distinguish the two kinds of *induction proper*, notice that two of these conclusions are far more certain than the first. We know that black swans have

---

[1] It would not be shorter if you had a broad river in the way.

'Cæteris paribus' means 'all other things being the same.' A good swimmer swims a mile in shorter time than a bad swimmer, 'all other things being the same,'—i.e. if they both have the same conditions of swimming. But 'cæteris' not 'paribus,' or let one tow a boat and the other not, and the bad swimmer may win the prize.

been discovered since the days of Socrates, but we feel convinced that bodies will fly towards the earth as long as the earth lasts, and explosions will follow the application of fire to powder as long as fire burns. The explanation of this difference is the secret of *scientific induction.* For a long time men were content with *inductio per simplicem enumerationem*, i.e., with saying 'all the swans we know or have heard of are white. ∴ All swans are white.'

Scientific induction begins with a demand for something more reliable as a guide to our action, and this something is found in the test of what is called cause or causation. Does the swan depend upon his whiteness for being a swan? Could he be a swan without being white? Yes; there are black swans. But could application of fire to powder be an application of fire to powder without an explosion? No, we should at once say, there is something wrong with the powder; it can't be true powder.

To borrow Mr. Fowler's exhausted pump instance. A guinea and a feather in an exhausted air-pump fall in equal times.

∴ These and all other bodies will, under the same circumstances, do likewise.

(2) Now here is an inference grounded on two assumptions.

1. That everything has a cause—(law of causation).
2. That like causes are followed by like effects—(law of uniformity of nature).

We know that in an exhausted air-pump there can be nothing to act upon this guinea and this feather unless it be the action of gravity, or that attraction to the centre of the earth which keeps all our feet upon the spinning globe though our heads may be downwards. We assume that everything has a cause. ∴ This fall in equal times must be due to a cause, and the only cause present is the action of gravity.

∴ The action of gravity is the cause of the fall.

Again, before inferring that the case will be the same with a shilling, a watch-key, a marble, &c., we must assume that like causes (e.g., the action of gravity here) are followed by like effects (e.g., falls of shilling, watch-key, or marble).

Hence, underlying every scientific induction are found these two assumptions or principles as already stated in a note (p. 163).

Thus scientific induction, not content with a mere enumeration of facts, founds upon the principle of causation some more reliable basis of operations.

(3) Imagine man in the infancy of the sciences (cf. pp. 9, 10), face to face with a tangled web of hopelessly complicated phenomena; causes and effects mixed myriads together. His innate love of order prompts him to arrange, and group, and classify; to put things together. He finds that some things always appear side by side, while others appear in succession, like always following like. As a student of scientific induction he will confine his attention to the sequences rather than the co-existences, as likely to produce more certain laws. He is not unarmed in this battle with nature. His instruments or weapons are *observation* and *experiment*. By the former he watches phenomena as they occur; by the latter he changes their surroundings and sees what happens then. Thus observation is natural; experiment artificial. The astronomer *observes* the movements of the stars and makes laws; the botanist *observes* the growth of plants and makes laws; and these laws are certain in proportion as they express the relations between cause and effect. But the plurality of causes (or the fact that the same effect may be produced by several causes, e.g., death by all kinds of things), and the intermixture of effects (or the fact that effects may be produced partly by one thing, partly by another, as water is produced by the mixture of two gases), baffle mere observation, and experiment is called in to 'isolate the phenomena,' as in the air-pump the action of gravity is isolated. By experiment we can produce in our laboratories

the lightning in the shape of electrical sparks without the danger or difficulty of watching it in the storm. To find the effect of a given cause, experiment is better than observation ; but to find the cause of a given effect, observation is better than experiment.[1]

It is easy to see that the more complicated the phenomena the more difficult it is to distribute and arrange right causes with right effects, and so to produce laws. In astronomy and geometry much has been done, and they are called deductive sciences because all their phenomena have been grouped and classified under a few simple laws. In chemistry (read carefully pp. 12, 13), it is very hard to isolate the phenomena. In physiology it is still more so. To reduce human action to laws of cause and effect in a social science,[2] is a hopeless task for the same reason. The ancient philosophers not receiving from their fathers the results of previous observation and experiment as we do, unaided by all the triumphs of modern invention, were compelled either to give up the attempt of making laws altogether, or else to make theories and force the facts into subjection to them, and call these theories laws. Now hypotheses (see p. 182) are assumptions to explain phenomena but only assumptions, and they only become laws by standing the test of verification. Given the phenomena of movements of the sun, winds, and rivers. The ancients said 'a priori' (p. 185), 'things that move are living beings. These move. ∴ These are living beings.' They spun out their theories and applied them to the facts, instead of starting with the facts and working upwards. As the spider spins his web, so they their theories—out of themselves; whereas they should have flown from fact to fact like the bee. 'Hypotheses non fingo,' cries a wise man. But hypotheses, like wine, must not be condemned because some people are intoxicated with them.

---

[1] This is important. Think it out and illustrate it by instances from your own experience.

[2] *Statistics* are 'observation systematised' for this purpose.

Many great discoveries have been preceded by several tentative hypotheses, which would not stand the test of verification. Hypotheses are valuable enough if only subjected to impartial and rigid verification. The following summary of Mr. Fowler's rules should be remembered with regard to observation and experiment :—

1. *Precision.* Exact time at which an event occurs should be noticed, &c. Hence all kinds of instruments— watches, thermometers, dials, &c.

2. *Waste no time with immaterial circumstances.* Set aside everything which you know to have nothing to do with the effect you are trying to find the cause of ; e.g., if you're a doctor and see your patient is worse, don't ask him if he feels better.

3. *Vary the circumstances of the phenomenon as much as possible.* You do not feel well at Oxford. This is the phenomenon. You say to yourself, ' Probable cause—wine.' But before you can be certain, you must take the same wine at several other places; e.g., Brighton, Scarborough, &c. ; and you will often find that the wine is not the cause, but the climate or some other thing. This rule especially applies to observation.

¹ 4. *Isolate the phenomenon.* This is a rule for experiment. You bathe in sea-water. Your head aches afterwards. It may be the cold or the salt that is the cause of your pain. Isolate the cold by bathing in fresh water, and you will find out which it is.

N.B. The word ' cause ' is loosely used for any of the circumstances which precede a phenomenon. The phenomena ('any things that appear or can be observed by the senses ') which go to form that tangled web above described may be regarded, supposing their orderly arrangement to be completed,

¹ Remember these four by the word VIEW. V vary, I isolate, E exact, W waste.

as rows of soldiers one behind the other, each front rank to the row behind becoming a rear rank to the row in front—in other words, each set of phenomena becoming antecedents to the set in front of them and consequents to the set behind.

Often many antecedents combine to produce a consequent, and strictly speaking all these antecedents taken together are the *cause* of the effect. We speak of ' pulling the trigger ' as the cause of the explosion of a gun, but it is only the *occasion* or last cause. No explosion could have taken place, had it not been for the powder, the cap, the barrel, &c. All these are 'causally connected,' or ' part-causes of the effect.' A *cause* may be defined as a ' necessary and indispensable antecedent.'

The above applies rather to physical or mechanical causes. Efficient causes are those where the movement starts, so to speak ; e.g., the movement of the steam-engine is caused by the revolution of the wheel, which is caused by the working of the piston-rod, which is caused by the elasticity of the steam, which is caused by the water heated by fire, which is caused by the match applied by man.

Now this last agent, man, might or might not have applied the match, but the rest of the causes had no choice about it. They passed on the impetus mechanically. They were physical causes. The man was an efficient cause. A great question—the freedom of the will—hinges upon this distinction. Whether man is the efficient or physical cause of his actions. If he is only a physical cause, a social science would be possible.

(4) *The Methods.* ' Felix qui potuit rerum cognoscere causas.' Man rules by obeying nature. By finding what effects come from what causes, he can saddle those causes and bridle them (so to speak), and make them his beasts of burden. But he must study them carefully first, or they will crush him to atoms. Inductive logic provides him with these methods of experimental enquiry, as deductive logic gives him the rules of the syllogism. To make these methods clearer we shall give the examples first.

I. *The Method of Agreement.* You are an inductive logician. The savages of some new island come to you and beseech you to discover the cause of the madness which is raging amongst them. You tell them to bring you instances. They bring a dozen madmen. You find these men have several points in common ; they have all of them always been nervous, delicate, and worn with anxiety ; among other things they have all been bitten by a dog. You cannot isolate your phenomena—you cannot separate the nervousness, &c., and the bite. You cannot be sure which is the cause of the madness. You bid them bring you other instances of the same madness, but differing in everything else except that they have this madness. They bring all kinds of instances ; mad horses, mad squirrels, &c.; and you find that the instances of the madness are so many and so various that they may be said to have only one circumstance in common—a bite; in fact, the only thing you can say of them all is that they have been bitten by a dog.

You conclude that the bite is the cause or the effect of the madness. But you know the bite came first, so it must be the cause. This put into formal language becomes : [1]—' If two or more instances of the phenomenon under investigation have only one circumstance in common, the circumstance in which all the instances agree is the cause or the effect of the given phenomenon.'

Here phenomenon = madness ; instances = men, horses, squirrels, &c.; circumstances in common = bite.

The principle of this is that whatever can be cut out without affecting the phenomenon is not causally connected with it.

II. *The Method of Difference.* You have two guns of precisely the same value and workmanship. One 'kicks,'[2] and the other does not. Your man tells you that the reason is that one's a badly made gun. But after a careful investigation

---

[1] From Mr. Mill's Logic.

[2] If you use this instance, for ' kick ' read ' recoil ' throughout.

you find that the two guns, the kicker and the non-kicker, are precisely the same except in one thing—have every circumstance in common except one—viz., that the non-kicker is cleaned daily and the kicker is not cleaned daily. You conclude that the want of daily cleaning is the cause of the kicking of the gun. Put into formal language this becomes:—

'If an instance in which the phenomenon under investigation occurs, and an instance in which it does not occur, have every circumstance in common save one, that one being present only in the former, the circumstance in which alone the two instances differ is the effect, or the cause, or an indispensable part of the cause, of that phenomenon.'

Here the instance in which the phenomenon under investigation occurs is the 'kicking' gun; the instance in which it does not occur is the 'non-kicker;' the circumstance in which alone the two instances differ is the matter of cleaning; and the phenomenon is 'kicking.'

The principle of this is that whatever cannot be cut out without affecting the phenomenon is causally connected with it.

III. *The Joint Method of Agreement and Difference.* An officer who has seen much service finds that on certain days he is subject to twinges of rheumatism. Some of these days are fine, some wet, some cloudy, some clear, some in the summer, some in the winter, some in the town, some in the country; in fact these days of suffering have only one circumstance in common—an east wind. On the other hand the days when he feels no twinges are of all kinds also, they also differing in everything except one thing, *i.e.* having only one circumstance in common—the absence of an east wind. The officer concludes that the east wind is the cause of the twinges. This put formally becomes:—

'If two or more instances in which the phenomenon under investigation occurs have only one circumstance in common, while two or more instances in which the phenomenon does not occur have nothing in common save the absence of that circumstance, the circumstance in which alone the two sets of instances differ is the effect or the cause or some indispensable part of the cause of that phenomenon.'

Here the first set of instances are the wet, fine, cloudy, &c. days, *with* an east wind.

The second set are the various days *without* an east wind.

In the first set rheumatism (the phenomenon) occurs; in the second it does not. The first set have only one thing in common—an east wind; the second set only one thing in common—no east wind.

IV. *The Method of Concomitant Variations.* I rub two sticks together; the more I rub the hotter they grow, the less I rub the cooler they grow. Therefore the friction is the cause of the heat.

'Whatever phenomenon varies in any way whenever any other phenomenon varies in some particular way is either a cause or an effect of that phenomenon, or is connected with it through some fact of causation.'

V. There is also a *Method of Residues.* 'Subduct from any phenomenon such part as is known by previous inductions to be the effect of certain antecedents, and the residue of the phenomenon is the effect of the remaining antecedents;' e.g., A man is taken seriously ill after bathing and overtiring himself and drinking bad wine. Take away that part of his illness which is due to wine and the exhaustion, the residue is caused by the bathing.

N.B.—(i.) is the method for observation.

(ii.) is the method for experiment (the best of all the methods).

(iii.) for cases where you cannot isolate your phenomenon.

(iv.) for cases where the causes are *permanent.* We can modify heat[1] or the action of gravity, but so long as we are in the world we cannot exclude them.

---

[1] If we could banish the heat altogether, the sticks might be brought to that most useful method of difference. But many causes are perma-

(v.) most fertile in unexpected results.

Remember these methods by their instances. 'Even mad creatures recoil from twingey old officers—hot as two sticks rubbed together, and seriously ill from various causes.' Or by the hexameter and pentameter:

> Agreement, difference, first apart, then taken together,
> Concomitant variations—and a fifth—Residues.

FINIS.

Note.—The inductive syllogism runs as follows :—

This, that, and the other body fall to the ground.
This, that, and the other body are all bodies.
∴ All bodies fall to the ground.

This form is true enough, if we know *all* the instances. (*i.e.* in perfect induction). In imperfect induction our minor premiss is faulty, and our conclusion therefore not quite certain.

Archbishop Whately puts induction into a syllogism in Barbara by making a major premiss of the law of the uniformity of nature.

What | belongs to this, and that, and the other body | belongs to all.

Falling earthwards | belongs to this, that, and the other body.|

∴ Falling earthwards belongs to all.

Of course this is valid; the uncertainty of imperfect induction is, however, latent in the major premiss, which is only a generalisation from experience. Perfect induction seems to possess the certainty of deduction.

nent, and concomitant variations becomes a valuable method. The whole N.B. is important.

A LIST OF USEFUL FACTS IN LOGIC.

ABSOLUTE terms (opposed to relative terms) mean terms which are 'loosed from' (absolvo) any connection with other terms (*e.g.*, water); whereas relative terms have reference to (refero), or suggest other terms (*e.g.*, father [and son]—husband [and wife]). The pairs are called correlatives. You can think of 'water' by itself; but you can't think of 'father' without also thinking of 'son.'

ANALOGY.[1] An argument whereby from similarity between any two objects in points known we argue to further resemblance in points unknown, *e.g.*, moon and earth are two objects with known points of similarity (clouds, mountains, shape, &c.). The earth is inhabited; by analogy it follows that the moon is inhabited also. The value of analogy depends upon the number of points of resemblance known, as compared with the points of difference known, or the points of which nothing is known: *e.g.*, given two men; only known point of resemblance is—'both barristers.' One we know succeeds; but it would be a weak argument of an analogy to say ∴ 'the other succeeds also.' But given two fast men, with very many points of resemblance. One we know repents in after life. It would be a strong argument of analogy to say, ' ∴ the other repents.'

A PRIORI. } (See Method.)
A POSTERIORI.

ARGUMENTUM AD JUDICIUM means 'an appeal to the common sense of mankind.' Argumentum ad ignorantiam, 'an argument founded on the ignorance of adversaries.' Argumentum ad verecundiam, 'an appeal to our respect for some great authority.' Argumentum a concesso, 'a proof derived from a proposition already

---

[1] For 'analogous, equivocal, and univocal' words see p. 150-1.

conceded.' Argumentum a fortiori,[1]='arguing that you are right in a case which is *stronger* and better than one in which you were already allowed to be right.'

ATTRIBUTE (See Metaphysics).

AXIOM. A proposition accepted on its own merits, so to speak, which does not require proving itself, but from which we can prove other proposi tions : *e.g.*, ' The whole is greater than its part' (See Empiricism).

CATEGORIES. A list of 'summa genera ' (or ' highest classes ') given by Aristotle. Everything was said to belong to one or other of these genera, 'substance, quantity, quality, relation, action, passion, place, time, position, habit, or state.'

CAUSE means that without which a thing could not exist. Thus, there are four causes; remember them by a statue. Take away the stuff a statue is made of and there is no more statue; the material is called the *material* cause. Take away the shape and you destroy the statue; the shape or form = the formal cause. Let there be no maker, and the statue can't be made. The maker = the efficient cause. Let there be no aim or object in that maker's work, and the statue is a mass of confusion. The end or aim of the thing is its *final* cause. Hence there are four things without which a statue ceases to be a statue, and these are the four causes.

CONCEPTION (See Faculties).

CONCEPTUALIST (See Nominalist).

CONTRADICTORY (See Terms).

DIALECTIC. The art of discoursing. Also the old name of Logic. Several other meanings.

DICTUM DE OMNI ET NULLO. This means to say that what is true of a class is true of each individual in the class. Now to those who hold that classes are merely the sum of individuals composing them, this assertion is a truism. But to those who hold that there is something more in a class than the mere list of its individuals, it becomes an important truth.

DILEMMA. Remember the three instances of Dilemma thus : 'Science makes the Politician jest at Scripture (See Chapter on Hypothetical Syllogisms, p. 139).'

EMPIRICAL (ἐμπειρία, experience). Empirical knowledge is knowledge derived from experience. The knowledge of the old

---

[1] *E.g.*, even dogs can't digest nails, à *fortiori* dyspeptic invalids can't.

huntsman, or the old sailor is often empirical. They know what to do in each particular case, not because they have any principles or laws to act upon, but simply because they have an instinctive inclination to act in a certain way under certain circumstances. ' What do you do if a squall strikes the sail the wrong side ? ' you ask your sailor friend. ' Well, Sir,' he replies, ' I don't know what I does—I couldn't *tell* you—but when the squall comes I does it quite natural.' He has no principles consciously elaborated from the observation of particular facts, his knowledge is purely ' empirical.' A famous dye was lost not long ago by the death of the man who mixed the colours. ' Tell us,' said his employers, ' the principles upon which you mix, that the dye may not perish when you die.' ' Alas,' replied he, ' I know not how I do it. I can mix it myself, but I could not show another person how to mix it. The principles upon which he acted, the ' why ' and the ' wherefore ' and the ' how' he knew not; his knowledge was purely empirical.

EMPIRICISM is the technical name given to the theory that *all* our principles or laws are derived from experience. All men allow that *some* are. Such laws as ' the sun rises daily' are derived from experience or the observation of particular facts. But there are other laws which seem far deeper and more fundamental, such as ' The whole is greater than its part.' We can easily imagine a breach of the law, ' the sun rises daily ; ' but we can't *imagine* a part greater than its whole. Hence it is supposed that by nature certain laws are implanted in us to be apprehended by our reason and intellect, as particular facts are apprehended by our senses. Of this deep and fundamental character are the laws of thought. Nevertheless it is still true that in the science of thought we *start* with particulars ; for though the laws of thought already existed. we did not know them ; we *found* them, though we did not *make* them ; on our voyage of discovery we started with particulars and landed at laws (so to speak), though we did not make the laws any more than Columbus was making America when he started on his voyage of discovery. Empiricism would maintain that we make all our laws as we make all our laws of nature; *e.g.*, the laws of tides, sun, wind, &c., whereas others would maintain that laws, whereof the breach is inconceivable, are part of our nature, and found rather than made.

N 2

EXAMPLE. Analogy takes two instances as its groundwork. Induction takes more than two or several. Example takes one. By example we take one member of a class to represent the whole class, and argue from the individual to the class; *e.g.*, 'Tyrants are cruel.' Why ? ' Because Pisistratus (a tyrant) was cruel,' *i.e.* :

> P. was cruel.
> P. was a tyrant.
> .'. all tyrants are cruel.

The syntax of grammar abounds with instances. Care must be taken in this dangerous inference that the individual fairly represents the class.

EXPERIMENTUM CRUCIS. ' An experiment which decides between two rival theories, and shews which is to be adopted, as a finger-post shews which of the two roads is to be taken.' Crosses

are often erected in Roman Catholic countries at the cross-roads, or places where two or three roads meet. Hence also 'a crucial instance,' one which at once shows you which of two contending theories you are to adopt.

FACT. A word used in many senses, strictly means ' what is done or made.' Sometimes it is opposed to universal proposition, and sometimes to what is theoretical. Generally we speak of the facts of a case as the data (or ground, or material given), upon which we are to build up our inferences and theories. Facts would then mean all that is intuitively apprehended by sense or reason. And remember that at both ends of the domain of proof (traversed by the roads or paths called method) are ultimate and simple propositions which are called facts (see Method).

FACULTIES OF THINKING. *Sensation* is the faculty by which we are conscious of the presence of anything, without knowing anything except that our senses are affected (*e.g.*, child with dog, in ' Thought is Comparison '). When we are not only aware of the presence of a phenomenon, but also recognise it, we have attained to *perception*. *Conception* is the forming of concepts in the mind resulting from sensation and perception; it employs *imagi-*

nation (or the making of pictures or images in the mind). *Generalization* is a faculty by which we ascend to universals (ideas or propositions), which we apprehend by reason. By intuition, we know anything that we know without the aid of proof (see Method).

FIGURE OF SPEECH. A fallacy, which comes under amphibolia—*e.g.*, 'When we run we tread heavily upon what we run on; sometimes we run on empty stomachs; .·.sometimes we tread heavily upon empty stomachs.'

GENERALIZATION. (1.) Opposed to specialization (see Connotation of a Term).

(2.) Either the process or the result of our innate tendency to gather and group the like with the like.

GENERIC PROPERTY. 'That which belongs to the whole of a genus;' *e.g.*, 'hunger' is a generic property of 'man,' because it belongs to all the other species that go to make up the genus 'animal' to which the 'species' man belongs. Whereas 'cooks his food' is a specific property, for it belongs only to the species 'man.' Both are *properties*, for both follow from the connotation of man, 'rational animals.'

GRAMMAR is the science which is concerned with language primarily, whereas logic is only concerned secondarily with language. The grammatical analysis of the sentence, 'She stoops to conquer,' is into pronouns, verbs, &c., &c., but the logical analysis is into subject—copula—predicate. For ideas and not words are the important things to logic. Also rhetoric, or the science and art of persuading, differs from logic. For rhetoric persuades, logic convinces. In rhetoric, a fallacy which escaped notice would be no flaw, whereas in logic all fallacies are flaws. Rhetoric appeals to the feelings, logic to the intellect; and rhetoric relies upon the warmth of sympathy, while logic only recognises the cold, calm intellect. Hence rhetoric is the more popular, but logic the more true.

HYPOTHESES (or suppositions, ὑπὸ τίθημι, *sub pono*) are the conjectures hazarded by men who seek to establish laws from the observation of particular facts. If an hypothesis is right it becomes a law. I see several mad dogs at different times. I am anxious to establish a law as to the course of mad dogs. I hazard an hypothesis by saying, 'Happy thought! mad dogs run in circles.' A neighbour acts upon this theory, gets bitten, and dies. I reject this false hypothesis and select others, hoping to find one that accounts for all the particulars. I end with 'The course of mad dogs is a straight line if possible,' and many of my neighbours are saved by attention to this hypothesis, which passes into a law, or valid induction, as soon as it is found to stand the test of application to particulars. So Kepler rejected about twenty hypotheses before he found the true one as to the orbit or course of the planets; and the true hypothesis became at once a law.

INDIVIDUAL. The name given to that which is incapable of logical division, being a single object, as opposed to a class or group of objects.

INFERENCE. There are two kinds of immediate inference not yet mentioned. (i.) *Inference by added determinants*, which is of this form: 'The horse is a beast;.˙.the noble horse is a noble beast.' Care must be taken, for from 'The mouse is an animal' it does not follow that 'A huge mouse is a huge animal.' (ii.) *Inference by complex conception* of the form: 'The horse is a beast;.˙. the stride of the horse is the stride of a beast.' Care must be taken here also, for from 'Misers are men,' it does not follow that 'The most liberal of misers are the most liberal of men.'

INTENTION (for Intension see Connotation of Terms). A word is said to be of the first intention when it is the name of a thing; of the second intention when it is the name of the name of a thing. Thus, man, animal, &c., are names of things, while species, genus, &c., are the names we give to these names of things. We call 'animal' the 'genus,' 'man' the 'species,' and thus we name again (second intention) what were already the names (first intention) of things.

LAWS OF THOUGHT. The three great laws upon which all thinking is found ultimately to depend are :—

I. *The law of identity.* Whatever is, is. This we assume in every syllogism, for if one premiss changed while we were examining the other no conclusion could be drawn. In the old instance, when I seek to establish a connection between 'Socrates' and 'mortal,' after asserting 'Socrates is a man,' I may (so to speak) turn my back upon 'Socrates' in my attempt to find out how 'man and mortal' stand to one another, intending to return to 'Socrates' as soon as I have found this out. If 'Socrates' changes while my back is turned, what conclusion can I draw? On the contrary, by this law I know that if he *is* mortal he *is*, and I need not fear to leave that premiss while I busy myself with the other.

II. *The law of contradiction.* 'A thing cannot both be and not be.' A man can't be both tall and not tall at the same time. Of course, 'not tall' means everything that isn't called 'tall,' or you might say 'he is of middle height ',(which really means he is not tall). Upon this law, remember, opposition is based. Contradictory opposition would be powerless without this law. So also would reduction 'per impossibile.' (See Reduction of Figures, p. 133.)

III. *The law of excluded middle,* or that which excludes a middle state. 'A thing must either be or not be.' Upon this law depends dichotomy. (See Division.) N.B.—Keep distinct ideas of the laws of thought, the principle of the syllogism, the rules of the syllogism, and the canons of the figures.

METAPHYSICS. We divided all the universe into two parts. The ego, or person observing (subjective), and the non-ego, or objects observed (objective). We spoke of the attributes of these objects, as though each object was a something underlying its attributes, as our own souls and wills underlie our actions. To that underlying something (which we can't see, but imagine to be at the root of those attributes, as our own 'selves' are at the root of our actions) is given the name of *noumenon* (νούμινον, the thinking part), or substance as opposed to *phenomenon*[1](the part seen), or attribute. This mysterious something, this substratum, baffles all human powers of search. We think it lies somewhere behind the attributes, so we carefully remove the attributes one after another, and, wonderful to tell, when all the attributes are removed, there is nothing left at all! *e.g.,* a tree—take away all its attributes, 'shape, height, size, hardness,' &c., and you have nothing left. Still (by analogy from ourselves) we speak and think of objects as substance underlying attributes. Now, the science which is conversant with these inscrutable substances is metaphysics, or ontology.

NOMINALISTS, REALISTS, CONCEPTUALISTS. We saw that after a time the infant would start at the name of 'dog' without the

---

[1] From φαίνομαι, I show myself.

presence of a dog. Imagination enables us to picture to ourselves
' a dog ' which is not this or that dog, but simply a specimen possessing the necessary attributes and no more. Now, with regard
to this perfect specimen, this type dog, with all the essential but
none of the accidental qualities of ' dog,' the realist holds that it
actually exists (though out of our view) as a model or pattern
after which all dogs are fashioned, but to the perfection of which
type canine frailty cannot attain. The conceptualist says, ' Yes, it
exists, but only in our minds, by help of imagination. For seeing
many dogs, we gather into one imaginary dog all the essential
attributes, and this forms our conception " dog." ' The nominalist
says, ' That which we picture to ourselves as " dog " exists neither
in the mind nor out of it; it is merely " this or that dog "—some
particular dog we have seen, that we have photographed (so to
speak) in our souls. Thus the realists believe in the objective
existence of general notions (or common terms). The conceptualists in their subjective existence, and the nominalists allow them
no existence at all.'

METHOD ($\mu\acute{\epsilon}$ ' οδος). All our knowledge comes either from
intuition or from proof. There are the universals at the top, so to
speak, and the particulars at the bottom; and the first we apprehend by reason or intellect, the second by sense. Between these
two extremes, which we accept upon intuition, comes a group of
facts accepted upon proof or inference. Our ' whys ' and ' be-
causes' are obliged to stop when they reach the upper or lower
bounds, beyond which no proof can go. Now in the pursuit of
knowledge, as we pass through the domain of proof, we may travel
by two grand routes. We may either start with the universal
fact and journey southwards, or we may start with the particular
fact and journey northwards. And the roads or routes are called
$\mu\acute{\epsilon}\theta o\delta o\iota$ (ὄδος) or ' ways after something.' The journey downwards
is called the deductive, or the à priori, or the synthetic method,
and the upward journey the inductive, or the à posteriori, or the
analytical method. The former is also called the method of in-
struction, the latter the method of discovery. An instance of the
first would be proving that these two slate pencils couldn't enclose
a space by starting with the universal proposition, ' two straight lines
can't enclose a space.' (For this is one of the ' bounds ' of proof that
is accepted on intuition. If any one said, ' Why can't they ? ' you'd

answer, 'Because they can't, and there's an end of it;' for it is one
of the upper side 'bounds'). An instance of the second would
be to prove that 'all bodies are attracted to the centre of the
earth from such propositions as 'this book falls' (Suppose some
one said, 'Prove it falls' after you've seen it fall, you'd answer
'Can't you believe your eyes?' which means that it is a fact accepted upon intuition or sense—one of the lower-side 'bounds' of
proof). [As to Logic, in so far as we start with particulars to find
the laws, it's inductive, and in so far as those laws of thought
already existed, and we found and did not make them, it's deductive.] The truths at the upper end are said to be 'notiora
naturæ,' or better known in themselves and simpler; those at the
lower end are 'notiora nobis,' better known to us, but more complicated. Now as universal truths are simpler, so they are said to
be prior to particular truths. Hence à priori and à posteriori, though
generally we learn particulars first (as a child calls every man
papa). Smith is coming from Leicestershire to hunt with me.
A priori Smith is a 'good goer,' for 'all Leicestershire men are
good goers.' But Smith prefers the roads when he comes. A
posteriori, then, Smith is not a good goer. Synthesis is 'piecing
together' and analysis 'breaking up.' I am anxious to know
what this pudding is made of. I may either take simple elements, 'currants,' 'flour,' &c., and 'piece together' till I arrive at
the particular pudding in question, or I may send for Drug, the
chemist, and we may analyse the pudding till we get at the simple
elements of which it is composed. So with the proposition
'Socrates is mortal,' I may either take simple elementary propositions and piece them together as premisses until I arrive at this
conclusion, or I may take the proposition to start with, and eye it
narrowly until I find the simple elementary propositions of which
it is an instance. In one case I start with the universal, in the
other with the particular. Thus à priori and deductive and syn-
thetical may be regarded as the same, and à posteriori, analytical,
and inductive as their opposites.

OBJECTIVE (Metaphysics).

PRACTICAL UTILITY OF LOGIC. (1) As an abstruse study, it
braces the faculties of the mind. The harder the whetstone, the
keener the knife; and when you come to cut wood afterwards it's
easy. So it is with less abstruse questions after Logic.

(2) It exposes fallacies. These recur. The foxhunter, who
has seen many a fox drawn from a covert, is more likely to tell me
whether there will be a fox in yonder untried covert than a raw
novice, though neither may have seen the covert before. In
Logic the covert is a wordy speech, the fox the fallacy, and the
old foxhunter the cold logician. When the mob is on the point
of 'stones and fire' with delight at a speech, the logician is muttering 'accursed fallacy!'

(3) It is the study of the highest part of man, and man is the
highest thing in the world. Such a study must elevate and
ennoble us.

PROPOSITION. Indefinite propositions, where the quantity is
not specified, as 'girls are shy.' *Tautologous*, where the predicate
simply repeats the subject, and no information is given. *Eggs are
eggs.*

PSYCHOLOGY. The science of the whole inner nature of man;
i.e. of the whole mind or soul. Its object is to arrange and group
these phenomena and find their laws. One of its results is
that the mind or soul is broadly divided into three parts. The
purely intellectual, the purely animal, and the combination of
the two, or the union of reason and desire. Other sciences
then step in, take these three results, and produce further
improvements and arrangements. The science of ethics takes off
the union of reason and desire, as its subject-matter; and Logic in
the same way takes the purely intellectual part. Thus psychology
provides Logic with its material, as the art of making oars provides the art of rowing with material to work upon, or the science
of Euclid the science of land-measurement.

QUANTIFICATION OF PREDICATE. An attempt to quantify the
predicate as we do the subject. It is better not to adopt it:

(1) Because too troublesome.
(2)   „   too unusual a form.
(3)   „   makes one proposition into two.

RATIOCINATION. A grand name for syllogistic inference.
REALIST (See Nominalist).
RHETORIC (See Grammar).
SECUNDI ADJACENTIS. 'Brutus lives' is said to be a proposi-
tion secundi adjacentis (*i.e.* with copula lost in verb). Brutus
is brave, is tertii adjacentis with copula distinct.

SUBSTANCE (Metaphysics).

SYLLOGISM. Valuable as an inference, because it brings in truth in a new way, if it does not bring in new truth: moreover it is of the greatest assistance to the memory, as it dispenses with the recollection of innumerable particulars.

The *principle of syllogism* has these corollaries[1] (propositions following from itself, crowning it, so to speak):

(1) Terms whereof both agree with the same middle term agree with one another.

(2) Terms wherof one agrees and one does not agree with the same middle term do not agree with one another.

(3) Terms whereof neither agrees with the same middle term may or may not agree with one another.

TERMS (irregular). Categorematic terms are of the form 'house, horse.' Syncategorematic, 'of, to, from.' 'Grateful,' 'ungrateful' are positive and negative terms (presence and absence of a quality). 'Dead' is a privative term (loss of a quality). 'Less and greater' are opposites ('equal' intervenes). 'Negatives' are also called 'contradictories.'

TRUISM = a tautologous proposition; *e.g.*, whatever is, is.

This list Mr. Practical gave us as a bricklayer gives a series of dabs of mortar to fill up all crevices. He entreated us to read it over 'the night before the battle,' which I did with the zeal of a Roman Catholic soldier counting his beads, and the result was such as to exceed my wildest hopes. Both Dyver and myself passed with colours flying, Dyver being gently rebuked for giving up in a pass examination papers that would have carried off high honours, and I being complimented on the fact 'that I seemed to understand well what I wrote, and that it was refreshing to read papers that came from the writer's head; and were not a string of phrases from books learnt off, but never understood.'

The triumphant entry at home, the proud honours of a conquering hero, the affable condescension towards a world of inferiors who had never read Logic, or passed a public examination, the earnest desire of the female part of the family that the over-wrought brain that had soared to, and faced, and fought, and

---

[1] Corolla—a little crown.

mastered the mighty problems of that dimly awful monster Logic,
and passed through the anguish and tribulation and torture of an
Oxford examination, should have entire, absolute, and undisturbed
rest and enjoyment for months to come: all these things, as
likely to work upon the feelings, I leave to rhetoricians to de-
scribe; for my part, under the strong influence of my logical
teaching, I am disposed to cultivate the cold demeanour of the
logician, and to this end I habitually amuse myself by detecting
fallacies in the conversation of my father's friends, as they sip their
port after dinner, the exposition of which errors does not seem
so agreeable to them as it ought, though they are compelled to
acquiesce, my father nodding to them, as much as to say, 'Take
care, I expect he's right. He passed his examination very well in
Logic, you know.' For there is an undefined panic among them
at the very idea of Logic—as it seems to me, at least: and, failing
this, as a last resource I can always call any of their assertions 'A
flagrant instance of violation of the principles of constructive, conjunctive, hypothetical syllogisms,' and this produces a dead
silence at once.